ASTROPHOTOGRAPHY FOR THE AMATEUR

Michael A. Covington

Revised edition

CAMBRIDGE UNIVERSITY PRESS

Published by the Press Syndicate of the University of Cambridge
The Pitt Building, Trumpington Street, Cambridge CB2 1RP
40 West 20th Street, New York, NY 10011-4211, USA
10 Stamford Road, Oakleigh, Melbourne 3166, Australia

First published 1985
Reprinted 1986, 1987
First paperback edition (revised) 1991
Reprinted 1993, 1995

Printed in Great Britain at the University Press, Cambridge

British Library cataloguing in publication data
Covington, M.
Astrophotography for the amateur
1. Astronomical photography
1. Title
778.9′9523 QB121

Library of Congress cataloguing in publication data
Covington, Michael, A., 1957–
Basic astrophotography.

Includes index.
1. Astronomical photography. I. Title
QB121.C68 1985 522′.63 85–48025

ISBN 0 521 41305 2 hardback
ISBN 0 521 40984 5 paperback

GO

Soli Deo gloria

Contents

Preface

Amateur astronomy has undergone some striking changes in the past ten or twenty years. Space probes and high-technology observatories have made it harder for amateurs to make significant contributions to science, yet the popularity of astronomy as a hobby is soaring – and within it there have been shifts of interest. The amateur astronomer of two or three decades ago typically made his or her own telescope, from mirror-grinding to final assembly, and, when it was finished, embarked on a systematic study of the moon and planets, filling a notebook with drawings. Today's amateurs usually buy their telescopes ready-made, observe in a much more contemplative way, and want to take photographs of what they see. Astronomical photography is gaining in popularity even faster than amateur astronomy itself.

People's interests are not the only thing that has changed – the equipment and materials are different, too. Twenty years ago, catadioptric telescopes were a luxury for the wealthy; today, the prices have come down, and the Schmidt–Cassegrain and Maksutov have taken their place among the most popular amateur telescopes. The old familiar refractors and Newtonians are still around, of course, though they are likely to have shorter focal ratios than their predecessors, as well as accessories such as electric slow motions, compressor lenses, and even types of eyepieces that were unheard-of in the recent past. Quartz-controlled power supplies, cold cameras, Schmidt cameras, and gas-hypersensitized film are now 'off-the-shelf' commercial products.

This book is divided into three parts: 'Getting started', for the beginner, who may not have a telescope at all; 'Advanced techniques', for those with more experience; and 'Equipment and materials', a technical survey of cameras, films, and processing techniques. My goal is to give an introduction to astrophotography as it is done *today*, with the equipment and materials a not-too-advanced amateur can obtain and use. By 'astrophotography' I mean the photography of all types of celestial objects; the word is sometimes used more strictly to refer only to the photography of the stars.

I haven't covered everything. I've concentrated on 35-mm cameras and relatively small (20-centimeter or 8-inch and smaller) telescopes. Techniques that require exceptional technical skill, such as the use of cold cameras, are mentioned only briefly, with references to other sources of information. I have stayed away from video equipment, image intensifiers, and computer image processing; not because I think them unimportant, but because they are developing so rapidly that anything I could say today would very soon be out of date. On the other hand, I have felt free to be quite specific in describing conventional photographic materials because, although these will certainly change, the new products, when they appear, will be described by comparing them with the old.

The many amateurs who graciously allowed me to reproduce their work are acknowledged individually in the picture captions; they are of course only a sampling of the many skilled hobbyists doing astrophotography today. Several people deserve special thanks. Douglas Downing provided early encouragement. Bob Lucas helped me punch the chassis for the electronic guider described in Appendix D. In addition to contributing some fine photographs of his own, Dennis Milon, of *Sky and Telescope*, located many other contributors in the Boston area. The staff of *Astronomy* also helped me get in touch with contributors. The Eastman Kodak Company answered seemingly endless technical questions and supplied me with literature, along with permission to reproduce material.

Last but not least, in addition to suffering the usual privations that accompany being married to an author, my wife Melody drew most of the diagrams and line drawings, thereby giving a whole new meaning to the phrase 'in-house art department'.

This is also a good place to thank the many people who, at various times and places, helped me learn about astronomy and shared with me their appreciation for the sky. I owe much to the amateur astronomical communities of Valdosta, Georgia, where I grew up, and Albuquerque, New Mexico, where I spent an enjoyable summer in 1980.

I enjoy hearing from other hobbyists on all levels. Readers with questions, comments, or suggestions for revision are welcome to write to me care of the publisher, with the understanding that I cannot return photographs or other unsolicited material.

Michael Covington
Athens, Georgia

Preface to the revised edition

This revised edition (1991) reflects great advances in photographic film technology since 1985. We are winning the battle against reciprocity failure, and high-quality astrophotography is easier than ever before. The book has also been brought up to date in other ways, and minor errors have been corrected. Again I thank all who have written to me to share information, especially Ctein (of *Darkroom Photography* magazine). I also thank Melody, Cathy, and Sharon for their help.

Part I
GETTING STARTED

1
The stars

Stars and trails

To get started in astrophotography, all you really need is a camera that can make time exposures, a tripod or other steady support on which to mount it, and a cable release that will allow you to open the shutter and latch it open without vibrating the camera. Load your camera with color slide film rated at ISO (ASA) 200 or faster, go out on a starry, moonless night, aim the camera at a group of bright stars, open the lens to its widest *f*-stop, and make three time exposures – 20 seconds, 2 minutes, and 10 minutes.

When you get your film developed, look at the results. The 20-second exposure should show the stars much as they actually looked in the sky – though a careful examination will probably show that the film caught a few stars that were too faint to see with the unaided eye. In the 2-minute exposure, on the other hand, something will have happened. The star images will be short lines or curves instead of points – and in the 10-minute exposure, they will be longer lines or curves. It's almost as if the stars had been moving.

In reality, of course, it's the earth that moves; the earth's rotation is fast enough to create star trails instead of point images even on exposures as short as a minute or two. Stars directly above the earth's equator appear to move in straight lines; those in the northern or southern sky appear to trace circles

Fig. 1.1. *The familiar Big Dipper or Plough (Ursa Major) as seen in a 30-second fixed-camera exposure. 50-mm lens at f/1.8, Ektachrome 400 push-processed to ISO (ASA) 800. (By the author)*

Table 1.1. Fixed-camera exposure times for star fields (in seconds)

These values apply to 35-mm and 126 cameras. Cut them in half for half-frame 35-mm and 110; double them for 120, 220 and 620; and quadruple them for 4×5 sheet-film cameras.

Lens focal length (mm)	Declination of center of star field				
	0°	±30°	±45°	±60°	±75°
18	55	*65*	80	110	220
24	40	*50*	60	85	160
28	35	*40*	50	75	140
35	30	*33*	40	60	110
50	20	*23*	28	40	75
75	13	*15*	18	25	50
100	10	*12*	14	20	40
135	7.5	*8.5*	11	15	30
150	6.5	*7.5*	9	13	25
200	5.0	*5.5*	7	10	20
300	3.3	*3.8*	4.7	6.5	13
400	2.5	*3.0*	3.5	5.0	10

↑ Use this column if you do not know the declination.

Fig. 1.2. The Hyades and Pleiades, two large star clusters in Taurus. This 20-second fixed-camera exposure shows considerably more stars than were visible to the unaided eye. 50-mm lens at f/1.8, Ektachrome 400 pushed to ISO 800. (By the author)

Fig. 1.3. *The fishhook-shaped constellation Scorpius. Thirty-second fixed-camera exposure, 50-mm lens at f/1.8, Ektachrome 400 pushed to ISO 800. (By the author)*

Fig. 1.4. Star trails – this is what happens when the exposure is much longer than Table 1.1 recommends. Thirty-minute fixed-camera exposure of the constellation Bootes, 24-mm lens at f/4, Ektachrome 200 Professional. (By the author)

Fig. 1.5. Star trails centered on Polaris – but mostly sky fog from the lights of a nearby town. Thirty-minute fixed-camera exposure, 24-mm lens at f/4, Ektachrome 200 Professional. (By the author) Compare with Plate 1.1, which was taken in a more rural location.

around whichever of the two celestial poles is nearer. To see a dramatic illustration of this effect, aim your camera at Polaris and expose for perhaps two hours at *f*/8 – or look at Fig. 1.5 and Plate 1.1.

Another difference between the 2-minute and 10-minute exposures involves *sky fog* – the sky background looks much lighter in the longer exposure. The reason is that the night sky isn't perfectly black (it's usually deep blue or green), and the film picks up background haze that is too faint for your eye to see. The longer the exposure, the more sky fog the film records.

But as long as you're using a fixed tripod and want images that look like stars rather than trails, it's the motion of the earth rather than sky fog that limits how long you can expose. The practical limit for a particular picture depends on two things, the focal length of the lens and the position of the stars in the sky.

The focal length matters because a given amount of angular motion appears larger on the film with a telephoto than with a normal or wide-angle lens; hence longer lenses necessitate shorter exposures. The position of the stars also makes a difference because the nearer a star is to the celestial equator, the faster it appears to move. The distance from the equator to a star is called the star's *declination* and is expressed as an angle, like latitude on earth; stars with higher declinations move more slowly and allow longer exposures. You can find out the declination of a star (together with its *right ascension*, the celestial coordinate that corresponds to longitude on earth) from any star atlas.

Table 1.1 takes both of these factors into account and gives the longest practical exposures for a variety of situations. They are calculated with the formula:

time (in seconds) $= 1000/(F \cos d)$

where F is the focal length of the lens (in millimeters) and d is the declination of the star. This is for 35-mm and 126 cameras. For half-frame and 110-format cameras, the formula becomes:

$$\text{time (in seconds)} = 500/(F \cos d)$$

and for 6-cm formats (120, 220, and 620), it is:

$$\text{time (in seconds)} = 2000/(F \cos d)$$

Exposures for 4×5 cameras are twice as long as for 120.

If you don't know the declination, you can assume $\cos d$ is typically about 0.9 (or less if the star is in the far north or southern sky). But it would be a mistake to try to do calculations to improve on the accuracy of your first rough guess from Table 1.1; do test exposures instead. The tables and formulae are just rules of thumb, based on what, to my taste, constitutes a tolerable amount of elongation in star images; try longer and shorter exposures, and see how you like the results.

Films and lenses

I suggested using color slide film because it is least likely to meet with misfortune at the hands of the processors; slide film has only to be immersed in the right chemicals, in the right sequence, at the right temperatures, and out it comes, perfectly developed. The only thing that can go wrong is that your slides may be cut apart incorrectly – the black background of most astronomical photographs makes it very hard to tell where one picture ends the next begins. You can circumvent this by asking the processors to return your film uncut, and put it in slide mounts yourself.

When the photofinishers are required to make prints from negatives, the situation is different. The person or machine making the print has to decide how light or dark the picture should be, what adjustments to the color balance are needed, and whether the contrast is right. Very few photofinishers can be expected to make good prints of astronomical negatives, which look like nothing they've ever seen before. There's a good chance your star pictures will not even be recognized as printable negatives; most people do not want to pay for prints of pictures that didn't 'come out', and the photofinisher may conclude that all those dots and streaks mean only that something rather obscure has gone wrong with your camera.

That's why I say that if you don't do your own developing and printing, color slides are the best medium in which to work – you can rely on commercial processing. If you later want a print, you can tell the lab to 'make the print look as much like the slide as possible', which is generally a clear enough instruction. Of course, if you do your own darkroom work or work closely with a processing lab, you can use negative film and get prints made properly.

Either way, fixed-camera star photography requires a moderately fast film – at least ISO (ASA) 100, and the faster the better. Almost any fast film will work well; Ektachrome 200 and 400, Kodacolor Gold 1600, and Kodak T-Max 400 and P3200 are particularly good choices. Daylight-balanced color film is usually less susceptible to sky fog than film designed for tungsten light (though if your sky fog is brown or reddish, the opposite may be the case). Moreover, the film has to be fine-grained enough to keep the star images from disappearing into the grain. With the fastest films (ISO 1000 and higher) this usually happens to some extent, but with short exposures the advantages of fast film outweigh the disadvantages.

What about lenses? The situation here is rather complicated. Table 1.1 shows that wide-angle lenses with a short focal length allow the longest exposures; if exposure time were the only relevant

Fig. 1.6. *Part of the author's record-keeping system.*

factor, the shortest lenses would be the best. However, for reasons we'll go into in Chapter 7, the ability of a lens to photograph stars depends, not on its speed expressed in terms of *f*-stops, but on its actual diameter – and long telephoto lenses are the largest in diameter, if not the fastest. On the other hand, short lenses pack the star images together more densely on the film, resulting in a richer-looking picture. It would be possible to take all these factors into account and find the optimum mathematically except for one more thing – the film speed is not constant. Because of a phenomenon called reciprocity failure, most kinds of film become progressively less sensitive to light in longer exposures, and the extent of this speed loss varies from film to film.

My own experience has been that the camera's 'normal' lens – the one most often used for snapshots, for example, the 50-mm lens on 35-mm cameras – is usually the one that produces the best star pictures, but both wide-angle and moderate telephoto lenses have their uses. By all means experiment with all the lenses you have access to, and try exposures both longer and shorter than those given in Table 1.1.

Keeping records

In a sense, every astronomical photograph is a scientific experiment. It's impossible to say in advance how well a particular technique will work or in what ways it can be improved. This makes it vitally important to keep records of all the astronomical photographs you take. Otherwise, when you get good results you won't know how you did it; and when things don't turn out well, you will not be in a position to make modifications intelligently.

After experimenting with many ways of organizing records, I've concluded that strict chronological order is the best: any method that involves dividing things up by subject always results, for me at least, in something getting lost. I give each photograph a unique serial number consisting of the date in numeric form (year, month, day) followed by an identifying number for that date; for example, 83.03.24.06 would be the sixth exposure made on 1983 March 24. (If midnight, and hence a change of date, comes in the middle of a session, I ignore it and go on incrementing only the last two digits, so that a single session will have a single numbering sequence.)

In general, your records should include the following:

1 The object being photographed.

2 The condition of the air (clear or hazy, steady or unsteady, and so forth).

3 The details of the equipment used (remember to note the *f*-stop and include filters, if any). You'll probably want to devise standard abbreviations for commonly used setups.

4 The type of film, including notes on processing.

5 The date and time of the exposure. (From experience I've learned that these should be recorded in *local* time; conversions to GMT or the like should be done later. I once almost missed a lunar eclipse because in converting the time to GMT on the spot, I forgot to convert the date.)

6 The length of the exposure, in seconds.

Fig. 1.7. The constellation Cassiopeia, the Double Cluster in Perseus (to upper right of center), and other small star clusters. Thirty-second fixed-camera exposure, 50-mm lens at f/1.8, Ektachrome 400 pushed to ISO 800. (By the author)

Fig. 1.8. The Magellanic Clouds, satellite galaxies of the Milky Way, are not visible from most of the northern hemisphere. Akira Fujii captured them on film while vacationing in New Zealand. One-minute fixed-camera exposure, gas-hypersensitized Fujichrome 400, 24-mm lens at f/1.4.

Some advanced techniques

The main technical challenge with fixed-camera photography of star fields stems from the fact that most stars are faint; in all cases, you are dealing with an underexposed image. Any technique that increases either the speed of the film or the contrast of the image will improve the picture.

You can increase the effective speed of color slide film by 'push-processing' it – that is, by extending the time in the first developer. Many labs will do this for a small extra charge. The main disadvantage of push-processing is that it increases grain; trying to increase the speed by a factor of more than 2 to 4 makes the film so grainy that star images begin to disappear into the grain. Also, some films, such as Fujichrome 400, do not lend themselves to 'pushing'.

Another way to increase the speed of film is to *hypersensitize* it – that is, to lower the threshold that determines the minimum amount of light to which the film will respond. One way to do this is to *preflash* the film, by exposing it to a small amount of light in advance of the main exposure; just enough to produce a barely detectable amount of fog. Another is to treat the film with chemicals, of which the most popular is a mixture of nitrogen and hydrogen gases (see Chapter 7, and Fig. 1.8 and Plate 1.2). Gas hypersensitized film is often available commercially (one supplier is Lumicon Inc., whose address is given in Appendix A); for information on preflashing and other techniques, see *Increasing Film Speed*, by Mike Stensvold (Petersen, 1978).

Increasing the contrast of the image also helps. Color slides often benefits from being duplicated, since duplication almost always gives a contrast increase (Fig. 1.9 and Plate 1.3). One of the best slide-duplicating services is that offered by Kodak (Kodalux in the USA) through local photofinishers worldwide. Kodak duplicates are consistently good –

Fig. 1.9. A 20-second fixed-camera exposure of Orion through a 50-mm f/1.8 lens on Ektachrome 400, pushed to ISO (ASA) 800. (a) Original slide; (b) after duplication by Kodak. (By the author)

Fig. 1.10. Sagittarius, showing star clouds. Twenty seconds on T-Max 3200 film developed to ISO 3200 in T-Max developer. 50-mm lens at f/2. (By the author)

Fig. 1.11. The negative of Fig. 1.10. Note that more detail is visible than in the print.

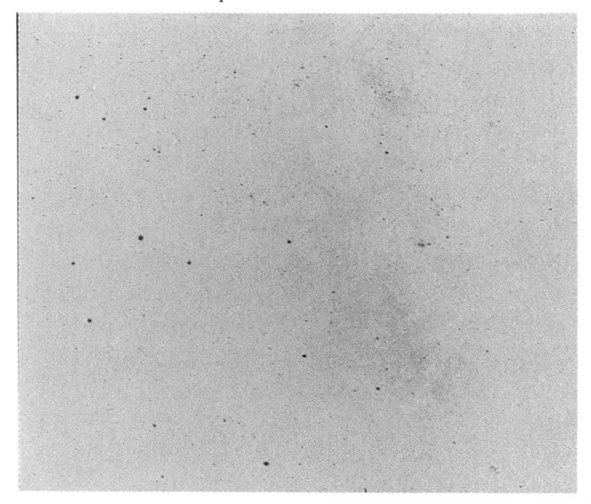

though duplication by Kodak no longer gives as much of a contrast increase as it did in the 1970s. (From the ordinary photographer's viewpoint, this is of course an improvement.)

Alternatively, you can do your own duplicating; see Chapter 10 for information. With fixed-camera star photographs, there is usually very little sky fog, and you want to make the duplicate lighter than the original in order to bring out detail. Try a wide variety of exposures, beginning with that indicated by your exposure meter and increasing it by two or three stops.

If you're using black-and-white film and don't have pictorial photographs on the same roll, give the film extended development in a high-energy developer. For example, develop any black-and-white film for 10 minutes in HC-110(A) at 20 °C (68 °F); or use one of the developers designed for push-processing, such as Kodak T-Max, Acufine, or Ilford Microphen, and extend the development time as much as the instructions allow. The result will be higher contrast and increased visibility of faint stars, which can be improved even further by treating the negatives with chromium intensifier. Then make the prints on No. 4 or No. 5 (Extra Hard or Ultra Hard) paper, making sure that the background is not so dark that faint stars disappear into it.

If you don't have access to a darkroom, you may want to send your black-and-white or color negative film to a custom lab or a lab that caters to amateur astronomers (check the advertisements in astronomy magazines). Another approach is to have your negatives developed locally and returned to you unprinted; you can then put them in slide mounts and view them as slides. Professional astronomers often view negatives instead of positives in order to be sure of seeing everything present on the original film; there is no reason why amateurs shouldn't use the same technique.

2
The moon

Lenses and image size

In many ways the moon is the easiest celestial object to photograph. It's practically the only thing in the sky whose brightness is so predictable that you can determine exposures accurately from an exposure table; under certain circumstances you can even use an exposure meter. Moreover, its surface shows such a wealth of detail that you can get pleasing, dramatic pictures of it with many different kinds of equipment.

To photograph the moon you need a long telephoto lens or a telescope of some kind; your camera's normal lens by itself won't give a large enough image to show any detail (Fig. 2.1). The size of the image of the moon on the film depends on the focal length of the lens, as expressed by the formula:

$$\text{image size (on film)} = \text{focal length} / 110$$

where the focal length is expressed in the same units as the image size (normally millimeters). As an approximation, just remember that the image size is a bit less than 1/100 of the focal length.

Table 2.1 shows the image size you'll get with lenses of various focal lengths, both on the film and on a ×15 enlargement (the highest degree of enlargement that is practical with most films and lenses – about the equivalent of a 16×20 inch print from a 35-mm negative, an 8×10 from a 110 negative, or a 30×30 from a 120 negative, if you were to enlarge the entire picture area).

Fig. 2.1. The moon over Mt Wilson Observatory as photographed by Jim Baumgardt. Because a 50-mm lens was used, the image of the moon is quite small.

Fig. 2.2. Through a long telephoto lens, the moon shows considerable surface detail but (usually) no craters. 400-mm Soligor lens at f/6.3, 1/500 second on Ektachrome 200. (By the author)

Fig. 2.3. The moon as photographed through a 400-mm telephoto lens, 1/500 second at f/6.3 on Kodak Technical Pan Film 2415 developed 8 minutes in HC-110(D) at 20 °C (68 °F) and enlarged about ×18. The crater Tycho is prominent in the white area at the top. (Melody and Michael Covington)

Let's suppose you're working with a 35-mm camera. If you use a 50-mm lens, the image of the moon on the film will be about half a millimeter in diameter, and even with maximum enlargement, you'll only get an image about 7 mm (a quarter of an inch) across on the print – far too tiny to show any detail. So you switch to your longest telephoto lens, a 400-mm. This gives you a 3.6-millimeter image of the moon that you can enlarge to about 55 mm (just over 2 inches) – big enough to show the lunar maria ('seas') and the face of the man in the moon, and perhaps some of the larger craters (Figs. 2.2 and 2.3).

If you then add a ×2 teleconverter, the effective focal length will be 800 mm, which is well within the useful range for photographing lunar detail; you can get an enlarged image of about 100 mm (4 inches) diameter, and, if your lenses are sharp enough and your tripod is steady enough, you may begin to see craters and mountains. A lens whose focal length is 600 mm or more by itself, without a converter, would of course be even better. Fig. 2.4 compares the effects of focal lengths of 600, 1500 and 2500 mm; as Plate 2.1 shows, a focal length of 1500 mm is quite ample for lunar photography on 35-mm film.

Table 2.1. Size of moon image with various focal lengths

Focal length	Size of image on film	Approximate size on ×15 enlargement	
(mm)	(mm)	(mm)	(inches)
28	0.25	3.8	1/8
50	0.45	6.8	1/4
100	0.91	14	1/2
200	1.8	27	1
300	2.7	41	1 5/8
400	3.6	54	2 1/8
500	4.5	68	2 5/8
600	5.4	81	3 1/8
750	6.8	102	4
900	8.2	123	4 3/4
1000	9.1	136	5 3/8
1200	10.9	164	6 3/8
1500	13.6	205	8
2000	18.2	273	10 3/4

Table 2.2. *Moon exposure table for ISO 400 film*
Times are given in seconds. For more complete exposure tables see Appendix B.

f/	Thin crescent	Wider crescent	Quarter	Gibbous	Full
2.8	1/500	1/1000	1/2000	–	–
4	1/250	1/500	1/1000	1/2000	–
5.6	1/125	1/250	1/500	1/1000	1/2000
8	1/60	1/125	1/250	1/500	1/1000
11	1/30	1/60	1/125	1/250	1/500
16	1/15	1/30	1/60	1/125	1/250
22	1/8	1/15	1/30	1/60	1/125
32	1/4	1/8	1/15	1/30	1/60
45	1/2	1/4	1/8	1/15	1/30
64	1	1/2	1/4	1/8	1/15

Table 2.3. *Longest exposure that will give sharp images without a clock drive*
These times apply to any celestial object.

Effective focal length range (mm)	For critical work (in seconds)	Where some blur is tolerable (in seconds)
90–180	2	8
180–350	1	4
350–700	1/2	2
700–1500	1/4	1
1500–3000	1/8	1/2
3000–6000	1/15	1/4
6000 and up	1/30	1/8

Fig. 2.4. *The moon as photographed with (a) a 600-mm telephoto lens; (b) a 32-cm (12.5-inch) telescope of 1500 mm focal length; (c) a 32-cm telescope of 2500 mm focal length. (Jim Baumgardt)*

a

b

c

Here are some important points to remember in photographing the moon through a telephoto lens:

1 Always use a tripod. You need as sharp an image as your camera can give, and you won't get maximum sharpness unless your camera is mounted on a *very steady* tripod. Use a cable release or self-timer to prevent the camera vibrating when you click the shutter. Needless to say, it is almost never possible to take acceptable lunar photographs with a hand-held camera, even when the exposure is quite short.

2 Set the lens at about *f*/5.6 or *f*/8 if possible, particularly if you are using a teleconverter. Most lenses are sharpest in this range.

3 Check the focus. If you're using a lens longer than 135 mm, or a teleconverter with any lens, you can't just set the focus to infinity and snap away; the lens may not actually be at infinity focus. The teleconverter introduces some errors of its own and magnifies any that are already present. The focus must always be checked at the viewfinder.

4 Remember that teleconverters multiply the *f*-ratio as well as the focal length. For example, a 100-mm lens set at *f*/8 becomes 200 mm at *f*/16 with a ×2 converter, or 300 mm at *f*/24 with a ×3. This has to be taken into account in calculating the exposure.

5 It's very hard to see or photograph craters on the full moon because the light is striking it so flatly. To capture the roughness of the moon's surface, photograph it when it is lit from the side, particularly at crescent or quarter phases.

6 If your lens gives only a small image of the moon, the way to get photographs or aesthetic value is to include other objects in the picture. For example, photograph the crescent moon shortly after sunset, while there is still enough light to show trees or buildings silhouetted against the horizon; if Venus or Jupiter is nearby, so much the better (Plate 2.2 and Fig. 2.6). This type of picture usually requires a much longer exposure than a picture of the moon by itself; so vary your exposure several stops on either side of the value you think will be correct (a procedure known as bracketing).

Determining exposures

I said earlier that you can sometimes use an exposure meter in taking pictures of the moon. This is true – but only if your camera has through-the-lens metering, and only if you are dealing with an effective focal length of 2000 mm or more, so that the moon more or less fills the whole picture. Otherwise, the meter will average the bright moon with the pitch-black background and the moon will be overexposed in the picture.

Ordinarily, then, you have to determine exposures from tables or by calculation. The basic formula for calculating any exposure is:

$$\text{time (in seconds)} = f^2/(A \times B)$$

where f is the *f*-stop to which the lens is set, A is the film speed expressed as an ISO (ASA) number, and B is a constant that indicates the brightness of the object being photographed: 10 for a thin crescent moon, 20 for a wide crescent, 40 for a half moon, 80 for a gibbous moon, or 200 for a full moon. For example, if you are taking a picture of the full moon at *f*/16 on ISO 400 film, plug the values into the formula as follows:

$$(16^2)/(400 \times 200) = 256/80\,000 = 0.0032 = 1/312 \text{ second}$$

Naturally, your camera won't have a 1/312-second setting; use the nearest that it does have (probably

Fig. 2.5. *The moon photographed through a 12.5-cm (5-inch) Schmidt-Cassegrain telescope used as a 1250-mm f/10 telephoto lens. Half-second exposure on Kodak Technical Pan Film developed for low contrast in Technidol LC; clock drive running. (By the author)*

Fig. 2.6. *The crescent moon, some stars, and Venus over Yale University. One second on Fujichrome 400 with a 100-mm lens at f/2.8. (By the author)*

1/250). Table 2.2 summarizes the exposure times for lunar photography on ISO 400 film; a more complete set of exposure tables is given in Appendix B. The calculated exposures are only approximations; variations in the transparency of the air and other factors can throw them off, so always bracket your exposures.

The same exposure formula, with appropriate values of *B*, applies to almost any picture-taking in the sky or on earth. It explains, among other things, why the standard *f*-numbers marked on most lenses are what they are. According to the formula, the exposure time is proportional, not to the *f*-ratio itself, but to the square of the *f*-ratio (f^2). The standard *f*-numbers and their squares are as follows:

f	1.4	2	2.8	4	5.6	8	11.3	16
f^2	2	4	8	16	32	64	128	256

You can easily see what is happening – whenever you go from one *f*-stop to the next higher one, you double f^2, and hence you double the required exposure time. The standard shutter speeds (1/1000, 1/500, 1/250, 1/125, etc.) are likewise arranged so that each one gives twice as much exposure as the one below. Whenever you change the lens opening by a particular number of stops, you can change the shutter speed by the same number of steps in the opposite direction to get an equivalent exposure. A 'one-stop' exposure change means that the amount of light or the exposure duration is cut in half or doubled.

There is one other thing to take into account: the earth is rotating, and if your camera or telescope doesn't have a clock drive to counteract its motion, there is a limit to how long an exposure you can make. The situation is the same as with fixed-camera star photography, except that now we want to eliminate trailing completely, not just keep it to a moderate level. The formula to use is:

longest practical exposure (in seconds) = $250/F$

where F is the focal length in millimeters. The moon's declination doesn't have to be taken into account because it doesn't vary enough to have a significant effect. Table 2.3 summarizes the results.

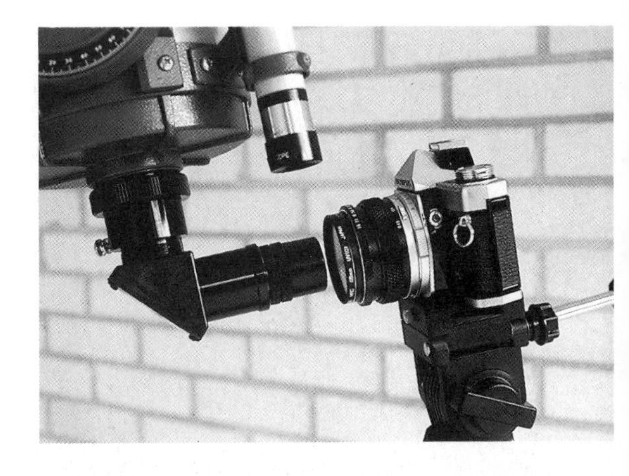

Fig. 2.7. With the afocal method, the camera and telescope can stand on separate tripods to minimize vibration.

Telescopes and binoculars

There are limits to the quality of the moon pictures you can take through even the best telephoto lenses, and you can usually get much better results by coupling your camera to a telescope. Some telescopes, particularly Schmidt–Cassegrains and Maksutovs, simply fit onto the camera body in place of a lens; this is known as *prime focus* photography. But in this section I concentrate on a different method of coupling, the *afocal method*, which can be used to link virtually any camera to any telescope without special equipment. (For other ways of coupling telescopes to cameras, see Chapter 5.)

In the afocal method, the camera takes the place of your eye at the eyepiece of the telescope; that is, you simply aim the camera (with its lens in place) into the eyepiece. No special adapter is needed; in fact, the best way to arrange the equipment is to put the camera and the telescope on separate tripods so that neither can transmit vibration to the other (Fig. 2.7). The camera should be equipped with a normal or medium telephoto lens, set to its widest opening. Take care to get the camera lens as close to the eyepiece and as well centered as possible.

If your camera is a single-lens reflex, you'll have no trouble focusing. Start with the telescope focused for normal viewing and the camera lens set in the middle of its focusing range; look in the viewfinder and focus, using the telescope's focusing knob first and then making fine adjustments by focusing the camera lens. If you happen to have a twin-lens reflex, focus with the viewing lens in front of the eyepiece, then move the camera so as to take the picture through the taking lens.

If your camera doesn't have through-the-lens viewing, you can still focus accurately – but it's a bit more complicated. In addition to the camera and main telescope, you'll need a small hand-held telescope of about 5 to 10 power; your main telescope's finder, or one side of a pair of binoculars, will do fine. First aim the hand-held telescope at the moon, or any other object more than about 150 meters (500 ft) distant, and focus it so that what you see is sharp. Next, aim the small telescope into the eyepiece of the main telescope and focus the main telescope so that you again see a sharp image. Then set your camera to infinity focus (the long-distance end of the focusing scale, usually marked ∞), put the camera in place, and make the exposure.

The hand-held telescope is necessary to reduce the amount of variation that the focusing muscle of your eye can introduce. When you focus your telescope for normal viewing, there is quite a range of settings that will give an image that appears equally sharp; you can put the image at any virtual (optical) distance from about 60 cm (2 feet) to infinity. The hand-held telescope ensures that you are placing the virtual image at infinity, so that a camera focused at infinity will be focused on it. (If you are over 50 years old and have perfect distance vision, you may be able to do without the hand-held telescope; the focusing range of your eye will be much less than a younger person's, and there is correspondingly less possible error. It's worth a try, at least.)

The effective focal length of an afocal setup is simply the focal length of the camera times the magnification of the telescope:

$$F = \begin{matrix}\text{focal length of} \\ \text{camera lens}\end{matrix} \times \begin{matrix}\text{magnification} \\ \text{of telescope}\end{matrix}$$

The *f*-ratio of the setup is the effective focal length divided by the diameter of the telescope objective:

$$f = F / \text{diameter}$$

Fig. 2.8. *A thin crescent moon photographed at the prime focus of a 12.5-cm (5-inch) Schmidt–Cassegrain telescope (1250 mm focal length, f/10). 1/15 second on Ektachrome 400. (By the author)*

Fig. 2.9. *Taken a few minutes after Fig. 2.8 with the same telescope and film, but using the afocal method with an 18-mm eyepiece on the telescope and a 100-mm telephoto lens on the camera, to give an effective focal length of 7000 mm at f/56. One second, clock drive running. (By the author)*

Throughout, I use capital *F* to stand for focal length and small *f* to stand for *f*-ratio. Be sure not to confuse the two.

To get good pictures of the whole of the moon's disk, you will probably need to use your camera with its normal lens and a telescope with a magnification of 20 to 50. (The magnification, in turn, equals the telescope focal length divided by the focal length of the eyepiece.) For example, one of the setups I've used in the past involves a 15-cm diameter (6-inch) Newtonian telescope of 1220 mm (48 inches) focal length, with a 32-mm eyepiece. This gives a magnification of 38 (that is, 1220 divided by 32), which in combination with a 50-mm camera lens works out to an effective focal length of 1900 mm (50 mm × 38) and an *f*-ratio of *f*/12.7 (1900 mm / 150 mm). The result is a 17-millimeter image of the moon on the film, enlargeable (in theory at least) to as much as 25 cm (10 inches) across.

Naturally, if you want to photograph a small area of the lunar surface, rather than the whole moon at once, you can use higher magnifications. I often use a 12.5-cm (5-inch) Schmidt–Cassegrain telescope at ×70 together with a 100-mm camera lens to get an effective focal length of 7000 mm at *f*/56; Fig. 2.9 shows a picture taken with this setup.

The afocal method also works well with smaller telescopes, spotting scopes, and binoculars. If you put a camera with a 50-mm lens behind one of the eyepieces of a pair of 7×35 binoculars, you get an effective focal length of 350 mm at *f*/10 – and you can use the other eyepiece for aiming. The moon image with such a setup is a bit small for photographing lunar detail, but fine for eclipses, as well as for taking pictures of birds, distant scenery, and the like. There are commercially available brackets for coupling cameras to binoculars, or you can make your own.

Fig. 2.10. *First quarter moon at the prime focus of a 12.5-cm (5-inch) Schmidt–Cassegrain telescope, 1250 mm focal length, f/10, 1/60 second through #15 yellow filter on Panatomic-X push-processed in Perfection XR-1. As is usual for higher-magnification lunar photographs, this one is oriented with south up. (By the author)*

Fig. 2.11. *A 'close-up' of the crater Copernicus, taken the same evening as Fig. 2.10, but by the afocal method with the telescope operating at ×140 and a 100-mm telephoto lens in the camera. One second (clock drive running) with same film and filter. Because the telescope included a star diagonal, this picture is a mirror image of the other, reversed top to bottom but not left to right. (By the author)*

Fig. 2.12. *The full moon requires high-contrast development and printing. A 15-cm (6-inch) f/8 telescope with 32-mm eyepiece coupled afocally to camera with 45-mm lens, 1/125 second on Tri-X developed 7 minutes at 22 °C (72 °F) in D-19 diluted 1:1, printed on Kodabromide No. 5 paper. In this case the moon is about to go into eclipse, as evidenced by a slight darkening at the lower right. (By the author)*

Fig. 2.13. *This impressive picture was taken by Dennis Milon at the prime focus of a 40-cm (16-inch) f/6 Newtonian telescope. The exposure was 1/30 second on Technical Pan 2415 developed in HC-110(D). The print was made on Polycontrast II Rapid RC paper with a No. 2 filter and developed in Selectol-Soft 1:1; the terminator area was heavily dodged.*

Films and processing

The moon is a relatively undemanding photographic subject; you can get pleasing pictures of it with practically any film. For this reason, it makes a good object to experiment on – you can try a variety of films and processing techniques and observe the different results while knowing that you'll probably get something usable no matter what happens. So this is as good a time as any to start getting acquainted with the factors that influence the choice of film for lunar and planetary photography.

The main technical challenge in photographing the moon is that, except when full, it spans a tremendous brightness range. Craters are most visible in the dimly lit areas near the terminator (the border between the illuminated and unilluminated areas of the moon), but it's almost impossible to expose such areas correctly without grossly overexposing the brighter areas. This means that you need a film with excellent exposure latitude (tolerance for overexposure and underexposure) and ability to hold shadow detail. At the same time, you need good contrast, particularly in the brighter areas, which are rather flatly illuminated – so the usual strategy of increasing the exposure latitude by reducing the contrast won't work very well.

What you need, in fact, is a film whose contrast increases with exposure, so that well-exposed areas show good contrast but the film's response to less well-exposed areas tapers off gradually, with gradually diminishing contrast. A film of this type is known as 'long-toed' because the curved 'toe' portion of its characteristic curve extends well up into the working range (see Chapter 9). Kodak Technical Pan Film developed in Technidol Liquid Developer, or T-Max 100 or 400 developed in HC-110, are particularly suitable. (T-Max films developed in T-Max developer may have excessively dense highlights; when in doubt, err on the side of underdevelopment.) Tri-X Pan is medium-toed; Plus-X Pan and the Agfapan black-and-white films are short-toed and not as suitable.

In general, you should develop the film to about the same level of contrast as for terrestrial photography, then print on contrasty paper (about No. 4) and do a fair bit of manipulation during enlargement (known as dodging and burning-in).

With color film, you more or less have to take what you can get. If you are able to make your own dodged prints, use a color negative film; if not, use one of the wider-latitude slide films, such as Kodachrome 64, Ektachrome 100 HC, or Ektachrome 400, since slides cover a wider density range than prints.

With the full moon, the situation is completely different. The full moon is a low-contrast subject because of the flat lighting. It requires a high contrast film such as Technical Pan (developed in HC-110 or D-19) or, for color slides, Fujichrome 400. If you decide to increase the contrast of regular film by overdeveloping, remember that overdevelopment increases grain; you can often get better results by developing the film normally and printing on high-contrast paper.

All this technical information about film needs to be counterbalanced by a few words about keeping things in perspective. The quality of your astronomical photographs depends largely on how familiar you are with *your* equipment and materials. This means that it's very important to pick one or two film-developer combinations (or at most, one for each general type of astrophotography) and stick with them until you know exactly what they can and cannot do. You will almost always get better results with materials you are thoroughly familiar with than with less familiar materials that are theoretically better suited to what you are doing.

3
Eclipses

Lunar eclipses

Many people first develop an interest in astronomy or astrophotography by watching an eclipse, and eclipses of the moon are the kind that people see most often, since they are visible from wide areas of the earth at once. An eclipse of the moon is caused by the moon passing into the earth's shadow, which blocks the sunlight that normally illuminates it; the darkening takes place on the moon, not the earth, so it is visible from everywhere that the moon is visible.

Fig. 3.1 shows that the shadow of the earth, like any shadow from a light source of appreciable size, consists of two parts: an inner, uniformly dark part called the *umbra*, and an outer, fuzzy, gray part called the *penumbra* that gets gradually lighter toward the edges. (The same is true of the shadow that your hand casts in sunlight or the light of a single light bulb; try it.) Fig. 3.2 shows the moon's passage through the earth's shadow as seen from earth; the moon enters first the penumbra, which causes it to dim somewhat, and then the umbra, which plunges it into almost complete darkness. If the only light reaching the moon were direct sunlight, the moon would be completely black and invisible while in the umbra, but in fact a certain amount of light gets around the earth, through the atmosphere, and bathes the moon in a dim, coppery glow. The brightness of this glow is unpredictable, since it depends on the distribution of clouds and other obstructions (such

Fig. 3.1. A lunar eclipse.

(a) As usually shown in textbooks:

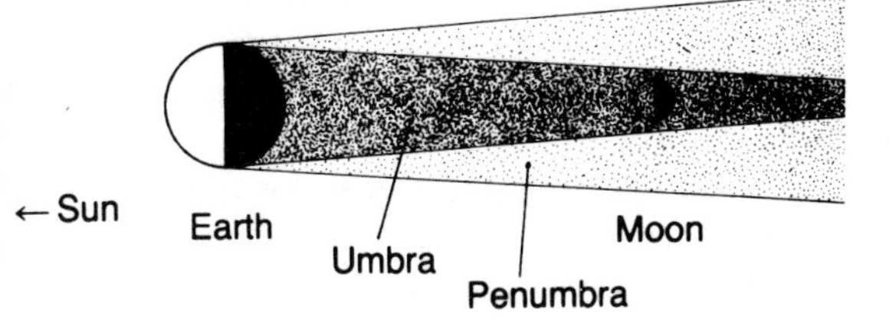

(b) A more accurate picture:

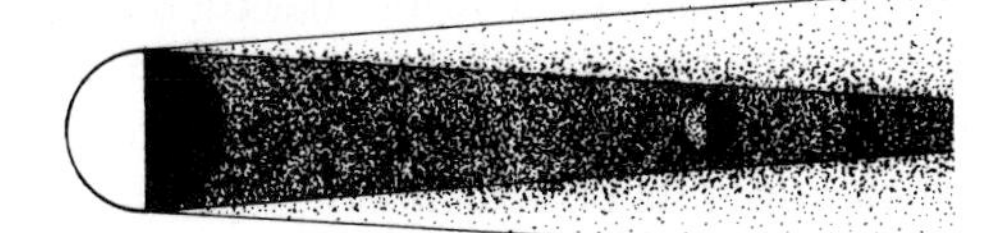

***Table 3.1.** Umbral lunar eclipses, 1991–2000*
All times are in GMT (UT).

Date	Partial or total	First contact	Mid-eclipse	Last contact
1991 Dec 21	Partial	10.01	10.33	11.05
1992 June 15	Partial	03.27	04.57	06.27
1992 Dec 9–10	Total	22.00	23.44	01.28
1993 June 4	Total	11.12	13.01	14.50
1993 Nov 29	Total	04.40	06.25	08.10
1994 May 25	Partial	02.39	03.31	04.23
1995 April 15	Partial	11.42	12.18	12.54
1996 April 3–4	Total	22.22	00.10	01.58
1996 Sept 27	Total	01.13	02.54	04.35
1997 March 24	Partial	02.59	04.40	06.21
1997 Sept 16	Total	17.08	18.46	20.24
1999 July 28	Partial	10.22	11.33	12.44
2000 Jan 21	Total	03.03	04.44	06.25
2000 July 16	Total	11.58	13.56	15.54

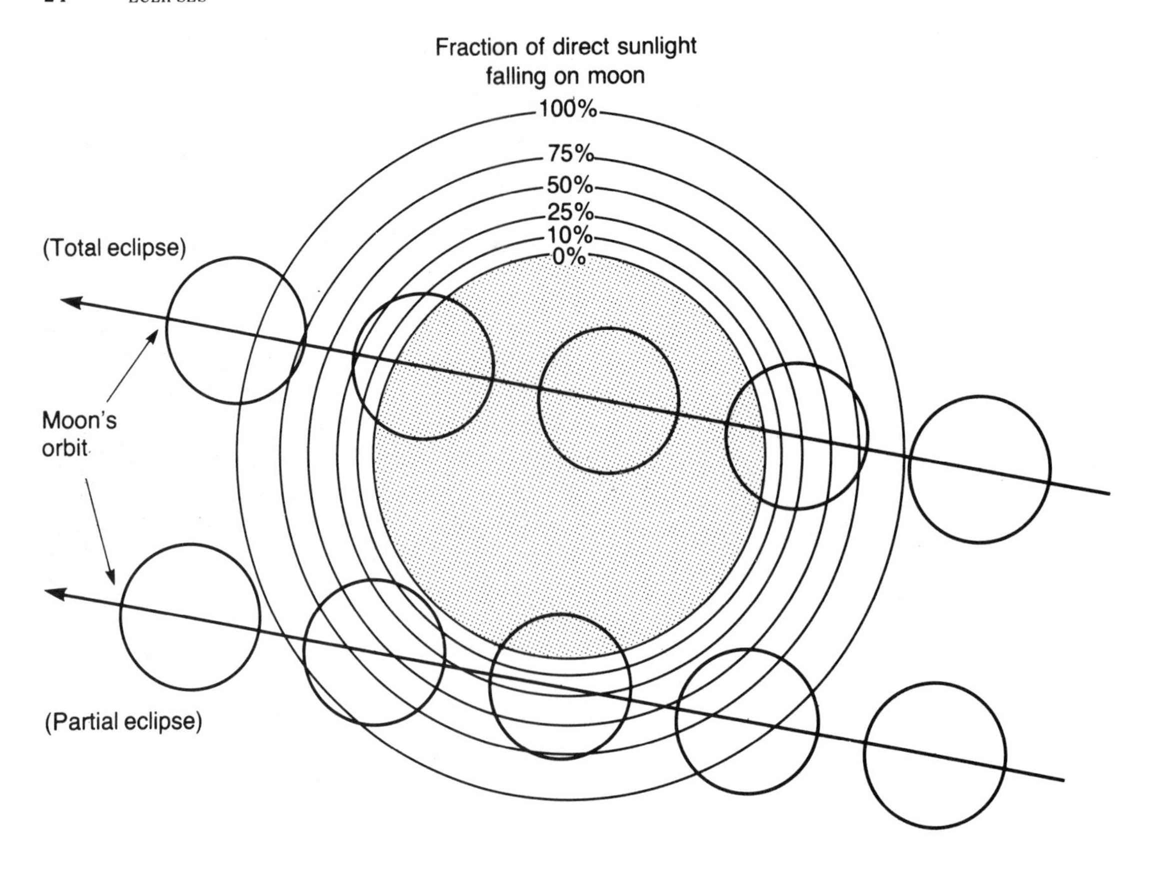

Fig. 3.2. Light intensity on the moon during a lunar eclipse, as seen from the earth. (The shaded area is the umbra.)

as volcanic dust) in the earth's atmosphere. Some scientists think fluorescence or phosphorescence of the moon's surface may also be involved.

Table 3.1 lists the lunar eclipses that will occur during the remainder of the twentieth century. Nearly two-thirds of these are *total* eclipses; that is, at some point during the eclipse the moon is completely within the umbra. The rest are *partial*, in that the moon grazes the umbra and is never more than partly immersed in it. There will also be frequent *penumbral* eclipses, in which the moon passes through some part of the penumbra without touching the umbra; these are not listed in the chart because the dimming is so slight as to be almost imperceptible.

To find out whether you can see a particular eclipse from your location, you need to know the time at which the eclipse takes place and whether the moon will be in the sky at your location at that time. The times in Table 3.1 are expressed in Greenwich Mean Time (also known as Universal Time, UT, or Coordinated Universal Time, UTC). To convert this to your local zone time, you'll have to add or subtract a fixed number of hours; Table 3.2 lists the conversions for most of the English-speaking world. If your time zone is not listed or you want to double-check your conversion, all you need is a shortwave radio, since all international broadcasters use GMT when they give the time of day. In addition, very accurate GMT signals can be obtained by telephoning 303 499-7111 in the USA.

To establish whether the moon will be in the sky at the time of the eclipse, all you need are the local times of sunrise and sunset. Since the moon is to be in the earth's shadow, the moon and sun will be in almost precisely opposite directions from the earth – which means that if the sun is in the sky, the moon isn't, and vice versa.

Let's put all this together with a concrete example. Suppose you want to observe the eclipse of 17 August 1989, and you live in California. Daylight saving time will be in effect in August, so your

Fig. 3.3. An example of what can be done even under adverse conditions and with relatively unsuitable equipment. This is the partial lunar eclipse of 17 July 1981, photographed by the author with a 100-mm telephoto lens at f/2.8 together with a ×3 teleconverter, exposing 1 second on Fujichrome 400, through a momentary break in the clouds.

Fig. 3.4. The lunar eclipse of 24 May 1975 as photographed by Douglas Downing at the prime focus of a 15-cm (6-inch) f/8 Newtonian. Three seconds on Fujichrome 100, 15 minutes after the end of totality. A small amount of the reddish-brown umbra is visible, along with the greatly overexposed penumbra.

Fig. 3.5. The lunar eclipse of 17 August 1970 photographed with a 15-cm (6-inch) f/8 Newtonian telescope and 32-mm eyepiece coupled afocally to a rangefinder camera with a 45-mm lens, and focused by the hand-telescope method. The exposure was 1/250 second, through high cirrostratus clouds, on Tri-X developed 7 minutes at 22 °C (72 °F) in D-19 1:1. (By the author, who was not quite 13 years old at the time)

Fig. 3.6. The same eclipse as Fig. 3.5, but photographed with a larger telescope through an unclouded sky. George East used Kodachrome II with a 25-cm (10-inch) f/6.4 telescope.

Fig. 3.7. The moon rising and coming out of eclipse on 30 December 1982. Akira Fujii made exposures at 5-minute intervals on Ektachrome 64 using a homebuilt 4×5-inch sheet film camera with a 400-mm lens at f/4.5, on a fixed tripod. The exposures varied from 1/2 second at f/5.6 for the thin crescent to 1/125 at f/11 for the nearly full moon.

conversion formula will be:

local zone time = GMT − 7 hours

The eclipse lasts from 01.21 to 04.55 GMT, and using the formula we find that this is equivalent to 18.21 to 21.55 Pacific Daylight Time the previous evening (that is, 6.21 p.m. to 9.55 p.m. on the 16th). The sun will still be in the sky at 6.21 p.m., so you'll miss the beginning of the eclipse, though the sun will set, and the moon will rise, well before the eclipse is over. Observers in New York (where local time = GMT − 4) will see the whole eclipse, which will last from 9.21 p.m. to 12.55 a.m. in their time zone; observers in England will see the eclipse begin at 2.21 a.m. local time, though the moon will have set before it ends at 5.55 a.m.

Once you've chosen an eclipse to observe, you'll have no trouble photographing it. The equipment needed to photograph lunar eclipses is the same as for photographing the moon generally; if anything, the requirements for sharpness are less stringent, since surface detail is of less interest (Figs. 3.3 – 3.6). You'll almost certainly want to photograph in color, in case the umbra turns out to be a striking copper hue (see Plate 3.1).

The main problem is that of determining exposure. During the partial phases of the eclipse, two approaches are possible: you can expose for the bright area outside the umbra, making the umbra itself look pitch-black, or you can try to get both the bright and dark parts of the moon into the film's sensitivity range. Appendix B gives exposure tables for both of these approaches; the second is considerably the more difficult of the two, since the umbra and penumbra differ in brightness by a factor of about ten thousand. The dimmest part of the penumbra is the area closest to the umbra, so the best conditions for success occur when the moon is almost completely inside the umbra and only a small sliver of penumbra remains. Naturally, it helps if the film has good exposure latitude.

Correct exposures for the totally eclipsed moon are unpredictable because the amount of light reaching the moon through the earth's atmosphere cannot be predicted in advance; some lunar eclipses are relatively bright, with the moon continuing to glow light orange even when eclipsed, while at other times the eclipsed moon is so dark as to be invisible even in a telescope. Two tables in Appendix B give suggested exposures for relatively light and relatively dark eclipses, but these are only approximations; to be on the safe side, bracket your exposures.

One thing you'll notice from the Tables is that exposures for photographing the totally eclipsed moon can be quite long – as much as several minutes at *f*/8, even on relatively fast film. This means that if you are using a relatively slow lens, you'll need to mount your camera on a telescope equipped with a clock drive to follow the earth's motion; otherwise, the limits in Table 2.3 still apply (though they can be stretched a bit if critical sharpness is not necessary). With exposures of a couple of minutes or more, the difference between solar and lunar drive rates can even become significant; the telescope will need to be driven slightly slower than its usual star-tracking rate because the moon is moving continuously in its orbit (see Chapter 6).

The dimness of the totally eclipsed moon does have one advantage; it makes it possible to take a picture of a star field with the moon in it, a feat that is normally impossible because the moon is too bright. Use a medium telephoto lens, mount the camera on a clock-driven telescope, and expose for two to five minutes; or use the fixed-camera technique described in Chapter 1.

If your camera can make multiple exposures, you can take a quite striking picture of the progress of

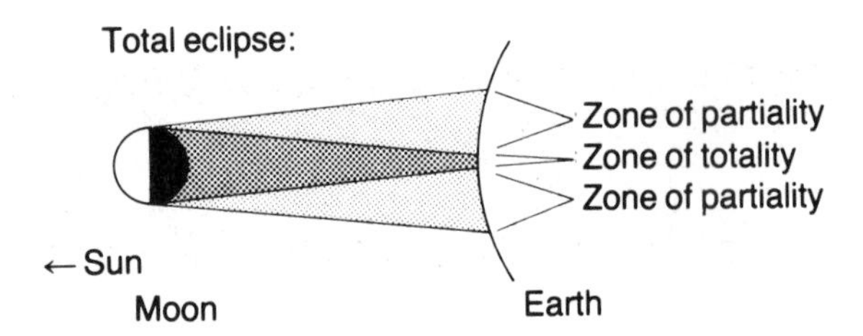

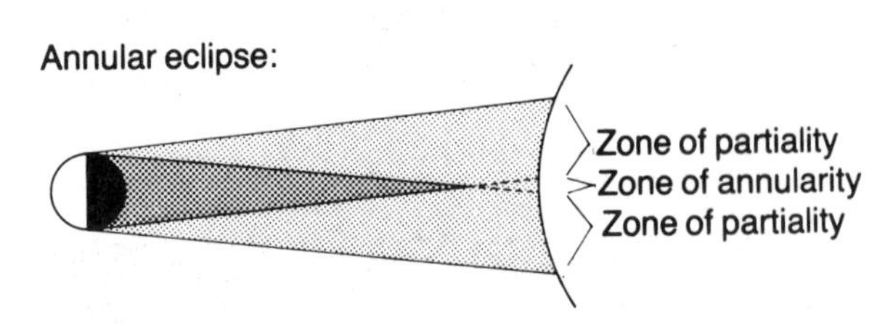

Fig. 3.8. Solar eclipse configurations in space.

the eclipse by placing the camera on a fixed tripod and exposing once every five minutes or so as the moon moves through the field (Fig. 3.7). Since the moon images will not overlap, each of them should be exposed normally. As an alternative, mount the camera on a clock-driven telescope, guide on a star (against which the earth's shadow is relatively stationary), and expose about once every 45 minutes; you'll get a photographic version of Fig. 3.2.

Table 3.2. *Converting GMT to local zone time*

Add or subtract the requisite number of hours, as indicated. If the answer comes out negative, add 24 and count it as the previous day (thus GMT − 4 for the eclipse at 01.21 on 17 August 1989 equals 21.21 on 16 August). If the answer comes out greater than 24, subtract 24 and count it as the following day (thus GMT + 2 for the eclipse at 22.22 on 3 April 1996 equals 00.22 on 4 April).

Time zone	Local zone time =	
	Winter (standard time)	Summer (daylight saving time)
Great Britain	GMT	GMT + 1
Continental Europe	GMT + 1	GMT + 2
South Africa	GMT + 2	GMT + 3
Australia (west coast)	GMT + 8	GMT + 9
Australia (east coast)	GMT + 10	GMT + 11
New Zealand	GMT + 12	GMT + 13
North America:		
Atlantic Time Zone	GMT − 4	GMT − 3
Eastern Time Zone	GMT − 5	GMT − 4
Central Time Zone	GMT − 6	GMT − 5
Mountain Time Zone	GMT − 7	GMT − 6
Pacific Time Zone	GMT − 8	GMT − 7
Alaska	GMT − 9	GMT − 8
Hawaii	GMT − 10	not used

Solar eclipses – partial and annular

Solar eclipses result from the moon passing in front of the sun – that is, from the moon's shadow falling on the earth – and the appearance of the eclipse depends on the observer's location relative to the shadow, which is of course moving all the time. A total solar eclipse is seen as total only from a small region at the center of the shadow – the *zone of totality* – which is a few kilometers or a few tens of kilometers in diameter and moves rapidly along a track several thousand kilometers long (Fig. 3.8). Surrounding it is the *zone of partiality*, a couple of thousand kilometers in diameter, within which the eclipse is seen as partial. This means that if you are situated on the path of totality, you begin by seeing a long, ever-deepening partial eclipse as the zone of partiality envelopes you. Then the zone of totality reaches your location and, if conditions are ideal, can take as much as seven or eight minutes to traverse it; during this time the bright disk of the sun is completely hidden from view and you see the corona. Suddenly the bright surface of the sun breaks through again; totality is past, and you are again in the zone of partiality. If you had been a few kilometers away, totality would have missed you altogether and you would have seen only a partial eclipse.

All this depends, of course, on the moon being

Total eclipse:

Annular eclipse:

Any eclipse viewed from outside the path of totality or annularity:

Fig. 3.9. *Solar eclipses as seen from earth.*

Table 3.3. *Forthcoming solar eclipses, to 1999*

This table lists all eclipses that will be visible (as total, annular, or partial) from any substantial part of the United States, Canada, or Western Europe. The paths described here are approximate; for more accurate information consult almanacs or astronomy magazines nearer the date of the eclipse concerned.

Date (GMT)	Total or annular	Path of totality or annularity
1991 July 11	Total	Pacific Ocean, Hawaii, Mexico, Central America, Colombia, Brazil (Exceptionally long duration of totality, over 6 minutes)
1992 Jan 4–5	Annular	Pacific Ocean, USA (touches land only at Los Angeles, just at sunset)
1994 May 10	Annular	Pacific Ocean, north-western Mexico, USA (New Mexico, Texas, Oklahoma, Kansas, Missouri, Illinois, Indiana, Ohio, Pennsylvania, New York, New England), Canada (Ontario, Maritime Provinces)
1998 Feb 26	Total	Pacific Ocean, Colombia, Venezuela, Atlantic Ocean
1999 Aug 11	Total	Atlantic Ocean (off Canada), England (Cornwall), France, Germany, Austria, Hungary, Rumania, Turkey, Iran, Pakistan, India

able to cover the whole sun as seen from earth. The moon is only barely big enough to do so; moreover, its distance from the earth varies, and when it is farther away than average (and therefore looks smaller) it can indeed fail to cover the whole sun. What happens then is that there is no zone of totality; an observer in the exact center of the moon's shadow sees the moon centered on the sun but not quite big enough to cover it, so that a ring-like area of the sun's surface remains visible (Fig. 3.9). This is called an *annular eclipse* (from Latin *annulus*, 'ring'); there will be a spectacular annular eclipse in the United States on 10 May 1994.

As far as observing techniques are concerned, of course, annular eclipses belong with partial eclipses, since what you observe in either case is the bright surface of the sun (the photosphere); moreover, any eclipse seen from outside the zone of totality or annularity is partial. This means that observing any solar eclipse involves viewing the photosphere – which is so bright that looking at it through a camera or telescope without proper eye protection will lead to eye injury and permanent blindness. This necessitates the following warning:

Never look at the sun through *any* optical instrument unless you are sure it is equipped with filters sufficient to reduce the light intensity to a safe level.

Fig. 3.10. *Shadow of the author holding a piece of paper with a ¼-inch hole in it during a partial solar eclipse. The image of the hole takes on the shape of the partially eclipsed sun. (Melody Covington)*

Obeying this warning is a matter of know-how, not just common sense, for two reasons. First, *this type of eye injury is usually painless*; you may not know at the time that you have suffered any permanent damage. What usually happens is that the injured person experiences a sensation of 'dazzle' exactly like the normal (and harmless) afterimage from a photographer's flash, except that instead of disappearing after a few minutes, it gets worse and within a few days develops into a permanent blind spot.

Second, and perhaps more important, the fact that the sun looks comfortably dim through a particular filter does not prove that the invisible wavelengths are being reduced to a safe level. The reason is that the eye cannot see all the wavelengths present in sunlight (light at wavelengths too short to see is called *ultraviolet*, and light at wavelengths too long to see, *infrared*). Photographic neutral-density filters and smoked glass are notorious for transmitting dangerous amounts of infrared; more about this later. Filters are safe only if instrumental tests have shown that they cut infrared and ultraviolet radiation by at least the same amount as visible light.

Around the time of a major solar eclipse, the news media often quote authoritative sources as saying that no type of filter is safe for direct viewing of the sun. Taken literally, this is not quite true; several types of filters (to be discussed below) have been shown to be safe when used properly, and astronomers have been using them for years without injury. What the authorities mean is that no type of filter is foolproof enough to be used by the general public without expert supervision. Safe solar viewing requires more than the average amount of willingness to follow instructions to the letter: the rules cannot be stretched even slightly.

When organizing viewing sessions for large groups, remember that people with no understanding of optics may sometimes misuse filters in ways that an experienced photographer would never dream of. At the May 1984 annular eclipse in Georgia, at least one schoolboy poked a hole in a filter in order to view the sun 'through' it, and several people thought aluminized mylar filters were made of aluminum foil. Others used insufficiently dense film filters without realizing that there was anything wrong with them, since they had never seen a filter of the correct density. Further, an annular or deep partial eclipse tempts people to look at the sun directly, without a filter. Even though most of the photosphere is covered, and the glare of the sun is much reduced, the uncovered regions are as hazardous as ever.

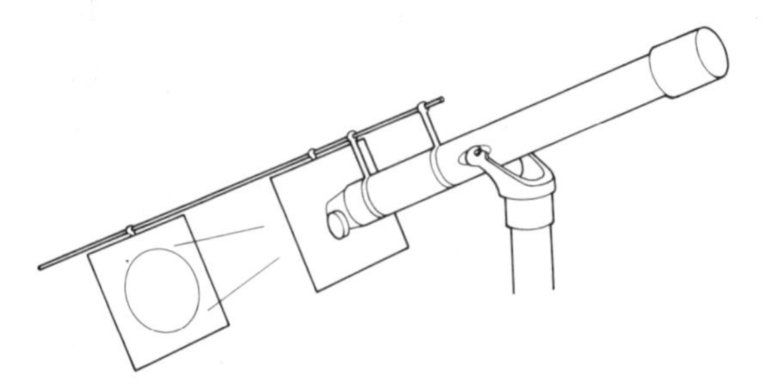

Fig. 3.11. *Projecting an image of the sun with a telescope. The first card, around the tube, casts a shadow so that the image on the second card is easy to see.*

Still, the often-heard statement that 'the only safe way to watch an eclipse is on television' is clearly an exaggeration: it is like saying that the only safe way to see the Grand Canyon is on a picture postcard (because you might fall in if you visit it in person). Eclipse watching can be done either dangerously or safely; it would be unfortunate to miss a spectacle of nature because of unrealistic fears.

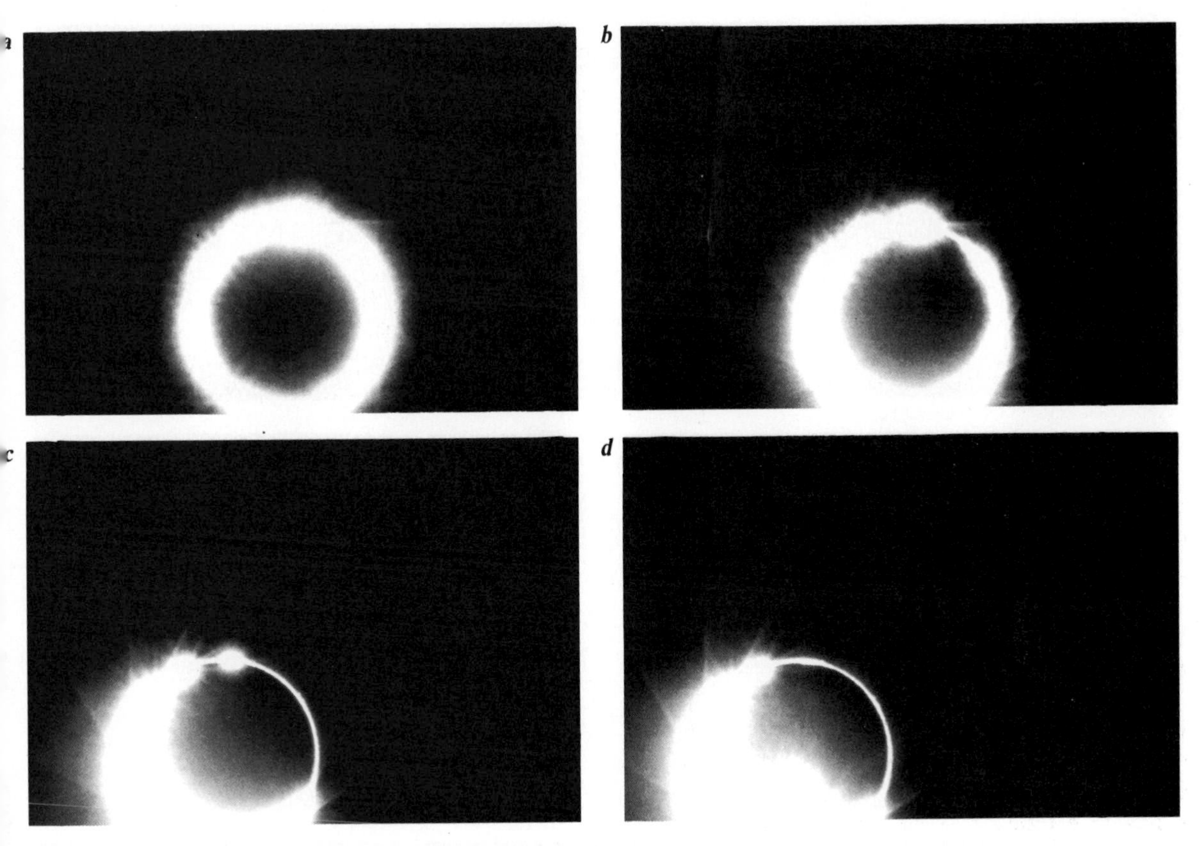

Fig. 3.12. The annular solar eclipse of 30 May 1984 photographed from Pendergrass, Georgia, by Melody Covington. (a) Mid-eclipse: the moon (black) is encircled by the greatly overexposed solar photosphere. (b) The moon has moved aside slightly, creating Baily's beads at the top. (c) and (d) Toward the upper right, the photosphere is completely covered and the chromosphere is visible. The thick spot in the middle of it is a prominence. Exposures about 1 second apart with a 400-mm lens at f/32, 1/1000 second, on Ektachrome 200, subsequently enlarged in slide duplicator. Because no filter was used, it was not possible to look through the viewfinder, and an attempt to aim the camera by its shadow was not wholly successful – the sun drifted out of the field.

In any event, you don't have to look directly at the sun; you can project the sun's image onto a screen and take pictures of the screen. The possibility of eye injury is then eliminated. The simplest instrument for projecting an image of the sun is a pinhole – either a hole punched in a card with a pin and used to cast an image on another card about 10 cm (4 inches) away, or a larger hole, casting a larger image on a larger screen. The hole-to-screen distance should be about 200 to 500 times the diameter of the hole in either case.

The principle involved is that, under such conditions, the spot of light that comes through the hole takes on the shape of the light source, not the hole. (You can demonstrate this by using square or triangular holes with the round, uneclipsed sun; if the screen is far enough from the hole, the image will be round.) If the hole is too big or too close to the screen, its shape influences the shape of the image, while if it is too small or too far away, the image is too dim and may be blurred by diffraction. A 7-mm (¼-inch) hole at 2 meters (6 ft), or a 25-mm (1-inch) hole at 7 meters (24 ft), is about right (Fig. 3.10). You can even observe a partial eclipse without any equipment, by making a tiny aperture with your thumb and index finger, and looking at its shape in the shadow on the ground.

The image that you get with pinhole projection is neither very sharp nor very bright, and you can get much better results by doing sun projection with a

Fig. 3.13. *The annular eclipse of 30 May 1984 photographed from Pendergrass, Georgia, with a 12.5-cm (5-inch) f/10 Schmidt–Cassegrain telescope and full-aperture Solar–Skreen filter. Each exposure is 1/30 second on Kodak Technical Pan Film 2415 developed 6 minutes in HC-110(D) at 20 °C (68 °F). At mid-eclipse, almost all the photosphere was covered. (By the author)*

4
5
6
10
11
12
16
17
18

Per cent transmission	Filter factor	Logarithmic density *(D)*
50	×2	0.3
25	×4	0.6
10	×10	1.0
1	×100	2.0
0.1	×1000	3.0
0.01	×10 000	4.0
0.001	×100 000	5.0
0.0001	×1 000 000	6.0

Table 3.4. *Three ways of measuring the light transmission of filters*

telescope (Fig. 3.11). Aim any telescope – or a pair of binoculars with one of the front lenscaps left on – at the sun, taking care not to look through it; hold a white card 10 or 15 centimeters behind the eyepiece, and focus until you get a sharp image of the sun (the edge should look crisp, and it should normally be easy to see one or two sunspots). This technique will generally give a good solar image, of adequate brightness, at least twice as large as the telescope objective; for instance, a 15-cm (6-inch) telescope will give a bright image 30 cm (a foot) across. The screen on which the image is projected will of course have to be shaded from direct sunlight; a convenient way to do this is to use a telescope that projects its image at a right angle, such as a Newtonian, or a refractor with a star diagonal, and put the screen at right angles to the sun. By doing this it is easy to display a 20-cm (8-inch) image on a slide projector screen, where several people can watch it. If you can project into a darkened room or tent, you can work with a dim but gigantic image a meter or so across.

Three precautions are in order. First, and most obviously, *make sure that no one tries to look into the eyepiece.* A good way to convince onlookers, especially young ones, that doing so would be unwise is to hold a piece of paper in the concentrated beam of sunlight emerging from the eyepiece; it will catch fire immediately. I have heard of a British astronomer who lights his pipe this way.

Second, note that the concentrated heat can damage eyepieces. Use an eyepiece that contains no cemented elements (a Ramsden or Huygenian), or an old or cheap eyepiece that you consider expendable, and allow it to cool down periodically by covering the front of the telescope or aiming it well away from the sun. (Do not aim it *slightly* away from the sun; the concentrated light and heat would then fall on the inside of the tube, possibly setting fire to it.) It is usually a good idea to mask the telescope's aperture to 7 cm (3 inches) or so; a square mask may be easier to make than a round one and is just as good.

Third, cover the finderscope so that you don't have any unpleasant encounters with the small solar image that it projects.

Once you have the sun's image projected onto a screen, of whatever size, it is a straightforward matter to take a picture of the screen with the image on it. Use an exposure meter as usual, whatever kind of film is handy, and remember that spots on the screen will not be distinguishable, in the picture, from spots on the sun.

Although projection setups are ideal for group viewing, they leave much to be desired as far as photography is concerned. The alternative, then, is to use a telephoto lens or telescope in combination with a protective filter. One of the most important characteristics of a sun filter is, of course, the extent to which it reduces the intensity of the light. As Table 3.4 shows, there are three ways of measuring this: as a percentage transmission, as a filter factor, and as a logarithmic density. These are related by the formulae:

$$\text{filter factor} = 100 / \text{percentage transmission}$$

$$\text{logarithmic density} = \log_{10} (\text{filter factor})$$

Note that when two filters are combined, the filter factors multiply, but the logarithmic densities add; this is the main convenience of using logarithmic densities. For example, a ×2 neutral density filter and a ×10 filter used together have a combined filter factor of 20 (that is, 2×10); expressing the same thing logarithmically, filters

Table 3.5. Safe and unsafe sun filters

Safe	Metallic film filters designed for solar viewing and used as directed (*best*)
	Two or three layers of fully exposed and developed conventional black-and-white film
	No. 14 welder's glass
Unsafe	Photographic neutral density filters of any density
	All other combinations of photographic filters, including crossed polarizers
	Filters made of color film
	Filters made of chromogenic ('silverless') black-and-white film (such as Ilford XP1 or Agfapan Vario-XL)
	Smoked glass
	Any filter through which you can see things other than the sun and very bright electric lights
	Any filter located near the telescope eyepiece, unless used with an unsilvered mirror or Herschel wedge
	Any filter not *known* to be safe

with densities of 1.0 and 0.3, used together, have a combined density of 1.3. Be sure not to confuse filter factors with logarithmic densities; a ×4 neutral density filter is a medium gray, while one with a density of 4.0 is so dark that nothing except the sun or a very bright light can be seen through it.

The surface of the sun is about 300 000 times as bright as an ordinary sunlit scene on earth; this means that in order to reduce its brightness to a comfortable level, you need a filter whose logarithmic density is about 5.5 ($=\log_{10} 300\ 000$). In practice, densities of about 4.0 to 6.0 are used (the lower densities on high-magnification telescopes and slow lenses that give an intrinsically dimmer image). The important thing is not so much the visual density as the density in the infrared and ultraviolet, which you can't see. A thorough set of safety tests for sun filters has been conducted by Dr B. Ralph Chou of the Optometry School at the University of Waterloo in Ontario. ('Safe Solar Filters', *Sky and Telescope*, August 1981, pp. 119–121; 'Protective Filters for Solar Observation', *Journal of the Royal Astronomical Society of Canada*, vol. 75, pp. 36–45, 1981.) Table 3.5 summarizes his results.

In particular, note that photographic neutral density filters (Wratten #96 or the like) are *not* safe to look through. This is a pity, since they are made in a wide range of accurately regulated densities. They are perfectly satisfactory for solar photography, of course, provided you can find some way of aiming and focusing your camera without looking at the sun through it.

The cheapest way to get a safe sun filter is to take a roll of conventional black-and-white film, unroll it in daylight or full room light so that it is fully exposed, then develop and fix it in the normal manner and use two or three layers of the resulting black film as a sun filter. A 120 roll of Verichrome Pan, which is very inexpensive, makes a piece of black film about 55 × 750 mm (2¼ × 30 inches); Kodak T-Max 100 and Technical Pan and Ilford Pan F and FP4 are also suitable. (You can develop the film in full room light, since it's fully exposed; or you can have it developed commercially.) Do not use color film or chromogenic ('silverless') black-and-white film.

The trouble with filters made of exposed film is that they are not very good optically; they don't give sharp images with lenses longer than about 200 mm in focal length. The same is true of welder's glass. But excellent optical quality combined with first-rate eye protection can be obtained by using a filter that consists of thin coatings of metal on glass or plastic. Of these, one of the best is the 'Solar–Skreen' marketed by Roger W. Tuthill (Box 1086, Mountainside, New Jersey 07092); it consists of two layers of Mylar plastic each of which has an aluminium coating on both sides, and it mounts in front of the telescope. Its only

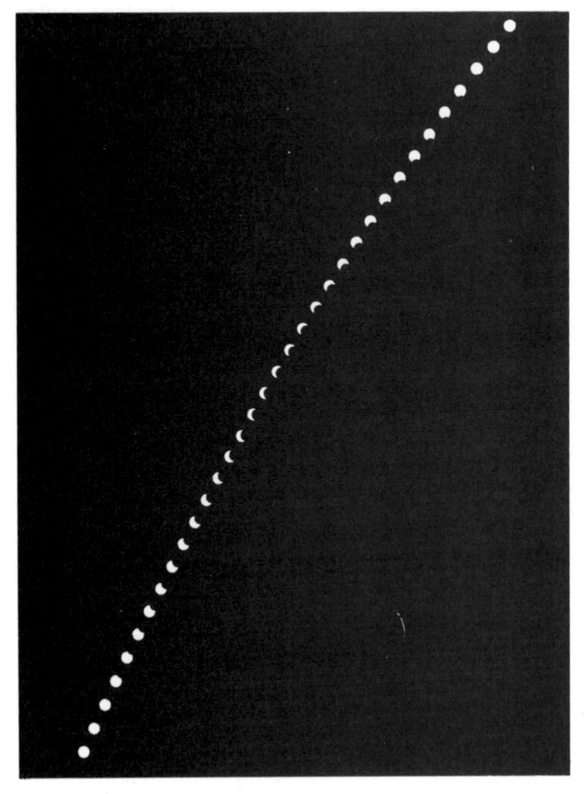

Fig. 3.14. Multiple exposures at 5-minute intervals of the solar eclipse of 31 July 1981, taken with a Mamiya C330 camera on a fixed tripod with a 100-mm lens at f/16 and a filter of logarithmic density 4.0. Each exposure was 1/60 second on Ektachrome 64. (Akira Fujii)

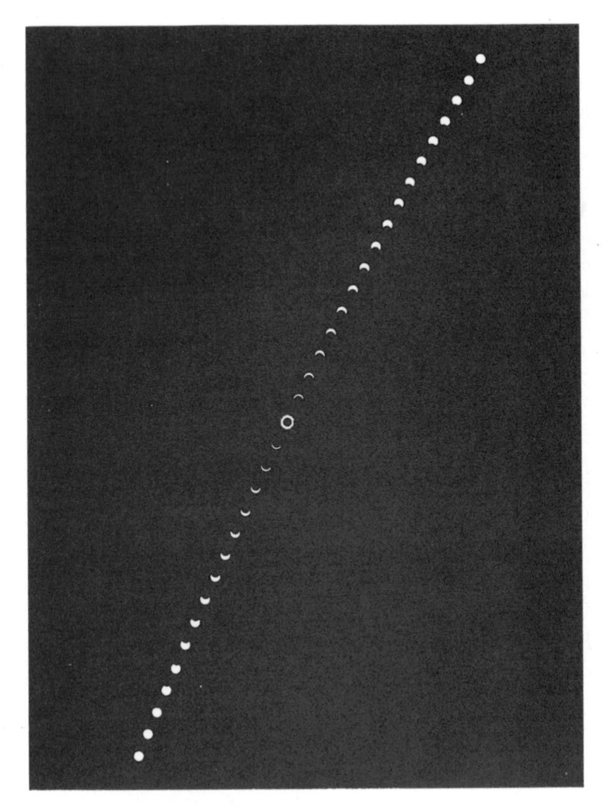

Fig. 3.15. The total solar eclipse of 19 February 1980, photographed from India with the same equipment, film, and exposure as Fig. 3.14 except that the exposure during totality was 1 second at f/5.6 with no filter. (Akira Fujii)

disadvantage is that it makes the sun look a peculiar electric blue color, but this is no real loss since the sun is a rather colorless object; for more realistic-looking color photographs, a #12 or #15 yellow filter can be added.

The importance of placing the sun filter in front of the telescope cannot be overstated. *A sun filter placed at or near the eyepiece is not safe*; the concentrated heat of the sun can crack and melt it, with disastrous results. This is true even if the telescope is quite small, since virtually all of the light that does not go into your eye has to be converted to heat within the filter. (Even giving the filter a shiny, reflective surface – as was commonly done a century ago – does not reduce heat absorption to a safe level.) The only safe way to use an eyepiece-mounted sun filter is to reflect the light off an unsilvered glass surface (a Herschel wedge) first – and, even then, you have to deal with image degradation resulting from heat waves in the air inside the telescope. It's better to keep full sunlight out of the telescope altogether by using a front-mounted filter.

Apart from filtration, the optical requirements for photographing the sun are the same as for photographing the moon – in fact, your eclipse-photography setup, minus the filter, can be tested on the full moon two weeks before the eclipse. The image sizes for the sun and the moon are exactly the same, and the time limits for exposures in Table 2.3 are equally applicable – though with the sun it is easy to keep exposures short by choosing filters of appropriate density.

Moreover, partial solar eclipses can be photographed on practically any kind of film. The exposure needed depends on the filtration used and can be determined in advance of the eclipse by practicing on the uneclipsed sun. In past years it was customary to take advantage of the sun's brightness and use extremely slow film for solar photography,

Fig. 3.16. *Different exposures show different amounts of corona. (a) 1/8 second on Ektachrome 200 at the prime focus of a 15-cm (6-inch) f/4 rich-field telescope; (b) 1/1000 second, same setup. The enlargements are to different scales, of course. Taken by Walter A. Singer at Ankola, India, on 16 February 1980.*

but this strikes me as a bad idea – if your filtration is so light that you can photograph the sun on films significantly slower than you would use without a filter for photographing the moon, then it probably isn't safe to look into the eyepiece.

There is one additional precaution that may not occur to you in advance, since the need for it arises only in the daytime: if you're using an afocal setup to photograph the sun, stray light has to be kept out of the space between the eyepiece and the camera lens. A convenient way to do this is to drape a piece of black cloth loosely around the camera and telescope after positioning them.

Solar eclipses – total

The totally eclipsed sun is at once an easy and a difficult photographic target – easy, in that many different equipment configurations, films and exposures can give pleasing results, but difficult, in that no photograph can record all the coronal structure and color visible to the eye. The reason for this is that the corona is much brighter nearer the center than at the extremities. Your eye adjusts to this brightness variation automatically, but photographic film does not; if you expose for the outer corona, you overexpose the inner regions, whereas if you expose for the inner corona, you lose the outer parts completely (Fig. 3.16, Plate 3.2). This makes it difficult to photograph more than a small amount of the streamer-like structure that is so striking visually; but at the same time it ensures that any exposure within quite a wide range will be right for *some* part of the corona.

To photograph the corona you need a field of view somewhat wider, and hence a focal length somewhat shorter, than for the partial phases of the eclipse. The useful effective focal lengths with 35-mm film range from about 1500 mm, which covers about twice the diameter of the solar disk, down to 200 mm or so; even a 50-mm 'normal' lens can be useful in recording the outermost parts of the corona. Suggested exposures are given in Appendix B. It goes without saying that, as no light from the bright photosphere can reach you, no filters are needed for viewing or photographing the sun during totality.

The best way to tackle the problem of the corona's brightness range is to use a color negative film and make, or get someone to make, a dodged print in which the relative density of the inner and outer parts of the picture is adjusted by hand. It is possible to construct a special dodging mask to get the right amount of correction more or less automatically; for two ways of doing so, see 'Some hints for photographers of total solar eclipses', *Sky and Telescope*, May 1973, pp. 322–6. The same article describes a way to use an occulting disk inside the camera to establish a density gradient with much the same effect. The unsharp masking technique described in Chapter 10 may also be helpful.

Among color print films, the Kodacolor Gold series is noted for its exceptional exposure latitude. Color slide films, though not quite as suitable, can also give good results; choose a relatively fast film with good exposure latitude, and stay away from high-contrast materials. Nowadays, few people photograph the corona in black and white, but if you choose to do so, develop the film to slightly lower than normal contrast in a developer such as Microdol-X.

The light of the corona is partly polarized, and it is interesting to take several pictures through a polarizing filter, changing the orientation of the polarizer for each exposure. Another peculiarity of coronal light is the region of reddish color that sometimes appears at about twice the diameter of the photosphere; its existence was disputed until Charles W. Wyckoff and Peter R. Leavitt photographed it on a special ultra-wide-range emulsion originally developed for the Apollo astronauts (*Sky and Telescope*, August 1970, pp. 72–3). It is worth trying to photograph the red layer on conventional film, especially in view of the improvements in color film that have taken place since the early 1970s.

But the corona is not the only thing to take pictures of. At the beginning of totality, the thin silver photosphere that has been visible suddenly breaks up into a number of disconnected spots, called *Baily's beads*, which consist of light coming through the spaces between lunar mountains (Plate 3.4). Within a few seconds, all of the spots disappear except one. What remains is called the *diamond ring effect* – the single gleaming bright spot together with the ringlike appearance of the inner corona look rather like a diamond ring in a jeweler's advertisement.

A moment later, the photosphere is hidden completely and the corona has come fully into view. But where there was a thin silver of photosphere half a minute ago, there is now a thin, glowing, reddish silver of something else – the *chromosphere* (Plate 3.4), a layer of ionized gas that lies between the photosphere and the corona. The chromosphere will likewise disappear from view within a few seconds, so the opportunity to capture it on film must be seized quickly. Finally, there are the *prominences* (Plate 3.5), great fountains of gas that glow red like the chromosphere but extend upward into the inner corona, usually high enough to remain visible throughout totality. At the end of totality the chromosphere, the diamond ring effect, and Baily's beads reappear in succession on the opposite edge of the moon.

The tables in Appendix B give suggested exposures for prominences; these should do equally well for the chromosphere. Exposures for Baily's beads and the diamond ring effect are hard to estimate in advance; use a slightly shorter exposure than for prominences, and hope for the best. Remember that the time to photograph Baily's beads and the diamond ring effect is at the *beginning* of totality, while you can safely continue looking into the eyepiece; when you see the first spot of photosphere appear at the end, take your eye away from the viewfinder and put the sun filters back on quickly.

There are also interesting things to see on the ground. Beginning about a minute before totality, and continuing until totality actually begins, the ground is covered with fast-moving parallel *shadow bands* a few centimeters wide, which reappear for an equal length of time at the end of totality. The cause of shadow bands is unknown; they probably result from irregular refraction in the earth's atmosphere.

Shadow bands are exceptionally hard to photograph because of their rapid motion and low contrast (there may be only a 1% difference in brightness between the bright and dark bands) and because the overall light level is constantly changing. They are most easily seen and photographed on a bright white surface – ideally, a slide projector screen – which the sun's rays are striking as directly as possible. Use a fast film with good exposure latitude, such as Tri-X or one of the chromogenic black-and-white films (Ilford XP1 or Agfapan Vario-XL), and make the prints on high-contrast paper or increase the contrast by repeated copying. (Do not use a high-contrast film for the original exposure; it would not give enough exposure latitude.)

The correct exposure for shadow bands is hard to predict; a rough guess is 1/250 second at *f*/4 (or 1/500 at *f*/2) on ISO 400 film. An aperture-priority automatic-exposure camera, especially one (such as the Olympus OM-2) that measures the light level throughout the exposure, can prove very helpful, since it can follow the changing light levels automatically; set the lens at its widest opening and hope for the best. The scientific value of pictures taken with an automatic-exposure camera is increased if there is some way of determining what shutter speed the automatic mechanism has given you; one way of doing this is to include in the picture an object moving at a known speed, such as a rotating disk, and measure how much of a blur it leaves on the film. The shadow bands themselves move too fast to be visible in exposures of more than 1/125 second, and much shorter exposures, of the order of 1/1000, are preferable.

There are interesting effects visible in the sky and on the horizon during totality. A fixed-tripod exposure of a large area of the sky (as in Chapter 1) will record the outer corona, a few bright stars, and any planets or comets that may be visible. (It is quite possible for a bright comet to be discovered during an eclipse, having been too close to the sun to be visible under ordinary conditions.) The horizon may appear orange or maroon in color, and the rapid motion of the zone of totality – which you can see coming at you out of the distance – is awe-inspiring.

Totality is, of course, short; you are fortunate to have three or four minutes in which to take all your photographs. This means that things have to be planned carefully and practiced in advance; make several 'dry runs', without film, of the whole sequence of activities you plan to carry out during totality. It is best to plan on using only about 70% of the time you will have available; things will inevitably go wrong and slow you down.

Wearing sunglasses for the last half hour before totality will help your eyes to dark-adapt. A tape recorder, playing a tape you have prepared in advance, can be useful for timing; you can prepare a verbal countdown so that you don't have to look at your watch to find out how many seconds of totality are left. This can be supplemented with another tape recorder to record your comments so that no time is lost writing things down. A camera with motorized film advance can save some time in making the exposures.

But don't get so busy taking pictures that you neglect the visual aspects of the eclipse. The pictures you take will probably look very much like the pictures hundreds of other people are taking at the same time, but neither your pictures nor theirs will capture the visual glory of the delicate coronal streamers or the feel of the sudden onrush of the moon's shadow. You can look at photographs any time; you may see totality 'live' for only two or three minutes in your whole life.

4
Comets, meteors, aurorae, and space dust

Comets

Any comet visible to the unaided eye can be photographed with a camera on a fixed tripod, using the technique described in Chapter 1. The exposure can be as long as two or three minutes if necessary, since even a trailed picture is undeniably better than no picture at all, and comets are rather fuzzy objects to begin with (Plates 4.1 and 4.2).

The most famous bright comet is Halley's, the only bright comet whose orbit is known well enough that its return to the inner solar system, once every 76 years, can be predicted in advance. We knew long in advance that Comet Halley would be visible in the evening sky after sunset in mid-January 1986; in the morning sky before sunrise at the end of March; and in the evening sky again in late April. (Comets in real life, unlike those in cartoons, do not streak rapidly across the sky; their motion is apparent only over a period of hours or days, and they appear in roughly the same place in the sky for several days in succession.)

Fig. 4.1. *Comet West (1976). A one-minute exposure by Walter A. Singer on Fujichrome 100 using a 50-mm lens. (Aperture not recorded – probably f/1.8).*

Unfortunately, in 1986, Comet Halley did not pass as close to the earth as it has done on many of its previous visits, and therefore did not look as spectacular as it did in 1910. It was visible to the unaided eye only under dark country skies.

But other comets equally bright or brighter turn up every few years; the only difference is that their orbital periods are so long – thousands or millions of years – that there are no prior records of them and prediction is impossible. Comets Ikeya-Seki (1965), Bennett (1970), Kohoutek (1973), and West (1976) are examples; each was discovered only a few weeks before it reached maximum brightness. Ikeya-Seki and Bennett were spectacular naked-eye objects, while Kohoutek, though a fine sight in binoculars, was fainter than early predictions had led the public to expect.

If you're interested in comets, then, it pays to watch the comet sections of magazines such as *Sky and Telescope,* or even to join the comet section of the British Astronomical Association, or the Association of Lunar and Planetary Observers. The information in magazines is about two months old when it reaches you; many a bright comet has come and gone within two months of its discovery, hence the advantage of belonging to an organization that can forward information to you more quickly.

In either case, the relevant information consists of the comet's predicted position, expressed as right ascension and declination (for plotting on your star atlas), and its expected brightness, expressed in terms of *magnitude.*

The magnitude system, which applies to all celestial objects, goes back to an ancient Greek

Plate 1.1. *This 1-hour fixed-camera exposure, centered on Polaris, shows how the earth's rotation makes stars appear to circle the north celestial pole. Taken by George East with a 55-mm lens at f/4 on Fujichrome 100.*

Plate 1.2. *This remarkable 30-second fixed-camera exposure by Akira Fujii shows the center of our galaxy as seen directly overhead from New Zealand; it rivals hour-long guided exposures that require much more elaborate equipment. Gas-hypersensitized Fujichrome 400, 24-mm lens at f/1.4.*

Plate 1.3. *A 20-second fixed-camera exposure of Sagittarius with a 50-mm lens at f/1.8 on Ektachrome 400 pushed to ISO 800. Faint detail was brought out by duplicating onto Kodachrome 25 with two stops more exposure than the meter indicated. Note the numerous star clusters and nebulae. (By the author)*

Plate 2.1. *The dark side of the crescent moon is often visible because the earth reflects light onto it. Kerry Hurd took this picture at the prime focus of a 25-cm (10-inch) Newtonian reflector of 1500 mm focal length, with a clock drive. Five seconds on Fujichrome 400 at f/6.*

Plate 2.2. *The moon and Jupiter over New Mexico, just before the occultation on 15 July 1980. 1/2 second on Kodachrome 64 through a 100-mm lens at f/4. A few minutes later the planet disappeared from view as the moon moved in front of it. (By the author)*

Plate 3.1. *The ruddy glow of the umbra contrasts with the bright white of the penumbra in this photograph by Dale Lightfoot. Eclipse of 6 September 1979; 9 seconds on Kodachrome 64 at the prime focus of a 20-cm (8-inch) f/5 Newtonian.*

b

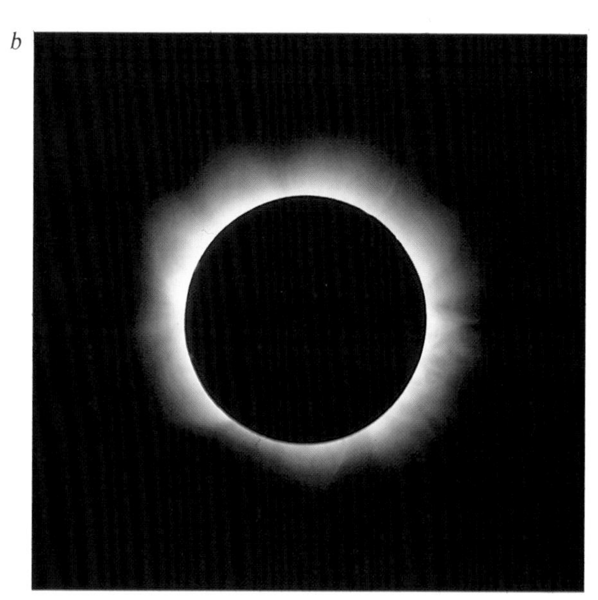

Plate 3.2. *The same eclipse as in Fig. 3.16, but photographed from Taita Hills, Kenya, by George East with an F/6.4 lens of 1000 mm focal length. (a) 1 second on Kodachrome 64; (b) 1/8 second.*

Plate 3.3. *As totality begins, the last visible sliver of the photosphere breaks up into Baily's beads (white), and the chromosphere (red) becomes visible. Taken by George East at Point Escuminak, New Brunswick, on 10 July 1972, with an f/12.4 lens of 1000 mm focal length. 1/250 second on Kodachrome II.*

Plate 3.4. *Here Baily's beads have disappeared but much of the chromosphere is still visible, along with the inner corona. Data as for Plate 3.3 but exposure of 1/125 second. (George East)*

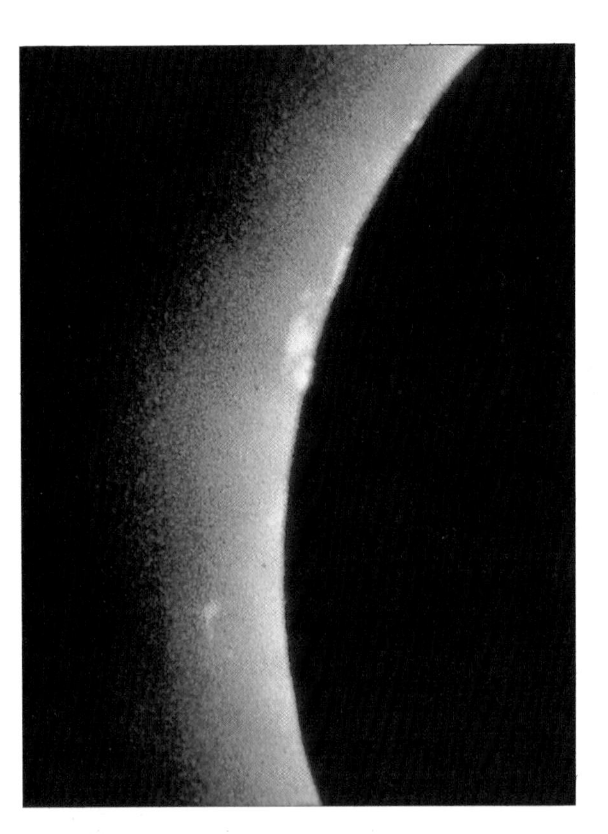

Plate 3.5. *A 'close-up' of solar prominences, one of which is detached from the chromosphere. Eclipse of 26 February 1979; negative projection with a 15-cm (6-inch) f/8 refractor. (Sherman Schultz)*

Plate 4.1. *Comet Ikeya-Seki (1965), photographed by Dennis Milon from the Catalina Mountains near Tucson, Arizona. High Speed Ektachrome in a 35-mm camera with a 50-mm lens at f/1.9.*

Plate 4.2. *Comet West (1976), photographed by Betty and Dennis Milon; a brief fixed-tripod exposure with a 45-mm lens at f/1.8. Note that slight trailing does not impair the effectiveness of the picture.*

a

b

Plate 4.3. *Two aurorae photographed by Sherman Schultz on High Speed Ektachrome. Both are 20-second exposures using a 50-mm lens at f/1.8.*

Fig. 4.2. *Comet Bennett (1970), photographed by Dennis Milon using a 4×5 Crown Graphic camera with Royal-X Pan sheet film and a 135-mm f/1.7 lens, wide open. The dark silhouettes are trees in the foreground that trailed during the 10-minute guided exposure.*

classification of stars as 'first class', 'second class', 'third class', and so forth, according to their brilliance; hence lower numbers designate brighter objects. The stars you see in the night sky range in magnitude from about 1 (the brightest) to 5 (or 6 in a dark country sky). With a 15-cm (6-inch) telescope you can see stars down to about twelfth magnitude. In modern times, the system has been extended in the other direction, through zero and into negative numbers, for brighter objects. Thus a few bright stars are now classed as magnitude 0, Sirius (the brightest star) is −1.4, Venus can reach −4, the full moon is −12, and the sun is −27.

Nebulous (fuzzy) objects, such as nebulae, galaxies, and comets, are generally a bit harder to see than stars of equal magnitude, especially if your eye is not trained for such work. A fourth-magnitude comet, for instance, may be hard to see without binoculars, even though fourth-magnitude stars are easy to see with the unaided eye. But you can photograph comets down to the fifth or sixth magnitude against a dark sky by using fixed-tripod exposures of a minute or less with a fast camera lens and fast film. Give the film as much push-processing as possible, and (in the case of black-and-white) print on high-contrast paper. Fainter comets, and better pictures of bright ones, require an equatorially mounted telescope; the techniques are those described in Chapter 7, except that you guide on the comet, which may be moving as fast as half a degree per hour relative to the stars.

Bright comets are always close to the sun, and you will often find yourself having to take pictures against a sky background that is not completely

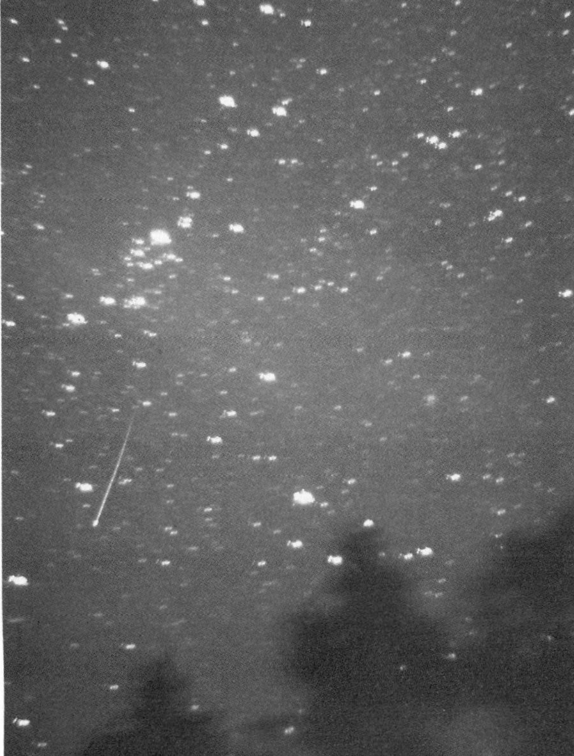

Fig. 4.3. Comet Halley as seen from Georgia on 23 March 1986. Two-minute 'piggy'-back' guided exposure, 100-mm lens as f/2.8, Focal (3M) 400 slide film, duplicated onto Ektachrome 100 to increase contrast. Globular cluster M55 is at lower left. (By the author)

Fig. 4.4. A Perseid meteor streaks through the constellation Perseus in this 5-minute exposure by Douglas Downing, taken on Fujichrome 100 with a 50-mm lens at F/1.8. The camera was mounted on the back of a telescope and guiding was done by hand.

dark. (Table 4.2, later in this chapter, gives the duration of twilight for various latitudes and times of year.) A bright yellow or orange filter (Wratten #12, #15, or #21) is often helpful in getting rid of such background light; in extreme cases, if you are making a guided exposure, use hypersensitized Kodak Technical Pan film and a deep red filter (#25 or even #29). Another approach is to use a film with a very short-toed characteristic curve, such as Agfapan 400 Professional, in order to reduce sensitivity to faint fog. It helps to try a wide variety of exposures, including some that you think will be too short, and to use contrast-increasing duplication techniques on the ones that turn out best.

Meteors

The way to photograph a meteor is to make a long exposure of a star field, with or without guiding, and hope that a meteor passes through the field while the shutter is open (Fig. 4.4).

Naturally, your chances of success will be greater if you make your attempt on a night when meteors are more common than usual – that is, during a *meteor shower*. 'Shower' is a relative term, of course; meteors never fall like hailstones, and a respectable shower is doing well to yield a visible meteor every two minutes. When there is no shower, the meteor rate is more like one every twenty or thirty minutes.

Meteor showers occur when the earth crosses paths with a swarm of meteoric particles that are circling the sun in a regular orbit; each meteor shower thus occurs at a particular point in the earth's orbit, and hence at a fixed time of year, and the meteors all appear to come from the same direction (the *radiant*, identified by the star of constellation for which the shower is named).

Table 4.1 lists the most important meteor showers; the hourly rate listed is for when the radiant is high overhead, and will be markedly lower if the radiant is near the horizon. Because the meteoric particles are not evenly distributed in space, there is always some variation in intensity from year to year. In the case of one shower, the Leonids (whose radiant is in the constellation Leo), this variation takes the form of approximately a 33-year periodicity: there was a dramatic Leonid shower in 1901, a respectable one in 1932, and a spectacular 'meteor storm' in 1966 in which the rate reached 40 meteors per second (Fig. 4.5) – but virtually nothing in the 'off' years. (Be prepared for the Leonids of 1998 or 1999; no one knows what to expect.) There are always more meteors after midnight than before, because after midnight you are on the forward side of the earth relative to its orbital motion.

In photographing meteors, use a normal or wide-angle lens, not only in order to cover a wide expanse of sky, but also because the small image scale helps to slow down the rapid motion of the meteors. As for film, meteor photography is peculiar in that you want a film that responds better to brief flashes of light than to faint, continuous glows – that is, a film with a high degree of *reciprocity failure* – so that you can catch as many meteors as possible without being bothered by sky fog. (This is the opposite of normal astrophotography, in which you need film that responds well to very long exposures.) The film of choice, therefore, is Kodak Tri-X Pan or Ilford HP5 Plus. Develop to high contrast. Color slide films such as Ektachrome 400 will work almost as well, and may even be more pleasing because the grain on the pictures is less prominent.

Exposure is tricky: the longer you expose, the more meteors you'll catch, but if sky fog blots out faint meteor trails, you lose what you've gained. To reduce sky fog, add a yellow or orange filter rather than stopping down the lens – it's better to reduce light selectivity than indiscriminately. And, of course, if you see a bright meteor go through the field you're photographing, end the exposure immediately – once you've caught a meteor, you don't need to accumulate any more sky fog.

Fig. 4.5. Dennis Milon made this 3½-minute exposure near the maximum of the spectacular 1966 Leonid shower, when the meteor rate briefly reached 40 per second. The two spots in the middle are meteors headed directly towards the camera (fortunately, they burned up before they could hit it). The bright stars are the Sickle of Leo. 105-mm lens at f/3.5, 120-format Tri-X Pan film developed 12 minutes at 19.5 °C (67 °F) in D-19.

Table 4.1. *Major meteor showers*

Date (every year)	Name	Typical max. hourly rate	Remarks
Jan 1–5	Quadrantids	100	The maximum, which lasts only a few hours, occurs on January 3 or 4. Radiant is near Theta Boötis
May 1–8	Eta Aquarids	20	Maximum around May 5
July 15–Aug 15	Delta Aquarids	35	Overlaps with Perseids. Maximum July 27/28
July 25–Aug 18	Perseids	65	The most dependable meteor shower; roughly the same every year. Maximum August 12
Oct 16–26	Orionids	30	Maximum October 21
Nov 15–19	Leonids	(varies widely)	Maximum November 17 The shower recurs in approximately a 33-year cycle and is due again in 1998 or 1999
Dec 7–15	Geminids	55	Maximum December 14

Aurorae

If you are fortunate enough to see a display of the *aurora borealis* (the northern lights), by all means photograph it. Use a normal or wide-angle lens, put the camera on a fixed tripod (the aurora does not share the stars' apparent motion), and expose for 10 seconds to a minute on fast film, with the lens wide open (Fig. 4.6, Plate 4.3). Photographs will often show color that was not visible to the unaided eye, since the eye does not perceive color in dim light. Also, if you do see color visually, the color in your photograph may not match it, because the light emitted by the aurora is confined to a few specific wavelengths, and one of these may fall on a particularly high or low point on the film's spectral response curve.

The visibility of the *aurora borealis* varies greatly from place to place; it depends roughly, but not exactly, on proximity to the north magnetic pole. In northern Alaska, Greenland, and Iceland, you will see aurorae on over 100 nights per year, twilight conditions permitting. The rate is more like 10 nights per year at Vancouver, Chicago, New York, or Manchester, and drops to an average of one night a year in Arizona, Texas, Florida, or France. Note that North America is much better placed for aurorae than Europe – in Rome the aurora is visible, on average, only once in ten years. (There is also a southern hemisphere aurora, the *aurora australis*, but none of the inhabited continents is well placed to see it.)

Aurorae are caused by electrically charged particles emitted by the sun, so they are most frequent when solar activity (of which sunsports are the most visible sign) is at a maximum. Solar activity varies in an 11-year cycle and reached maximum in 1990. It should hit minimum around 1997 and maximum again in 2001, give or take a year or so. The rise leading up to a maximum is quicker than the fall after one; conditions for observing aurorae remain good for two or three years after each peak.

Fig. 4.6. The Big Dipper forms a background for a spectacular aurora in this 20-second exposure with a 50-mm lens at f/1.8 on High Speed Ektachrome. (Sherman Schulz)

Dust in space

Among the promising but neglected targets for amateur astrophotographers are various glows in the sky caused by small particles of matter in interplanetary space. The easiest of these to observe is the *zodiacal light*, a glow that extends to the east and west of the sun along the ecliptic (the line in the sky that corresponds to the plane of the earth's orbit) and is visible after sunset and before sunrise (Fig. 4.7). The zodiacal light is a bit brighter than the Milky Way; it is caused by dust particles orbiting the sun, though the outermost parts of the sun's corona may also be involved.

The best way to confirm that you are looking at the zodiacal light, and not the last vestige of twilight (or the first sign of dawn), is to observe when the sun is far enough below the horizon that there is no twilight or dawn to be seen. The duration of astronomical twilight – that is, the length of time to wait after sunset in order to get a completely dark sky, or the length of time before sunrise that the sky begins to lighten – depends on your latitude and the time of year; Table 4.2 gives approximate values, and more extensive information can be obtained from the *Astronomical Almanac*. (The best way to determine the time of sunrise or sunset is by direct observation; the times given in newspaper are not always accurate, and the times given in the *Astronomical Almanac* have to be corrected for the difference between local mean time and standard time.) For observers in temperate latitudes, the ecliptic is most nearly perpendicular to the horizon, and hence the zodiacal light is easiest to see, in the spring evening sky or the autumn morning sky. (Phrased this way, this statement holds true for both the northern and southern hemispheres, though the calendar months involved are of course different.)

The zodiacal light should show up on exposure of between 20 seconds and 2 minutes at *f*/2 on ISO 400 film, provided the air is clear right down to the horizon (a situation most often achieved at sea, on

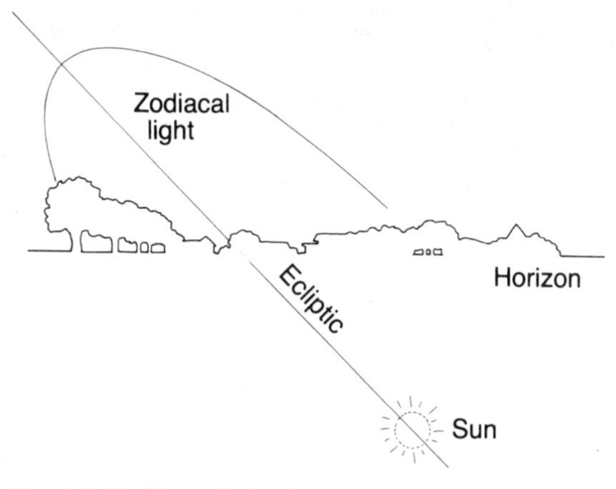

Fig. 4.7. *Unlike twilight, the zodiacal light is visible when the sun is more than 18° below the horizon.*

Table 4.2. *Length of twilight, in minutes*

Use the date scale corresponding to the hemisphere in which you are located. An asterisk indicates that the sky never gets completely dark at the specified latitude and time of year. Values are rounded to the nearest 5 minutes.

Date		Latitude (north or south)				
(northern hemi-sphere)	(southern hemi-sphere)	20°	30°	40°	50°	56°
Jan 1	July 1	80	90	100	120	140
Feb 1	Aug 1	75	80	90	115	130
March 1	Sept 1	75	80	90	110	125
April 1	Oct 1	75	80	95	115	140
May 1	Nov 1	75	85	105	140	*
June 1	Dec 1	85	95	115	*	*
July 1	Jan 1	85	95	120	*	*
Aug 1	Feb 1	80	90	110	150	*
Sept 1	March 1	75	85	95	120	145
Oct 1	April 1	75	80	90	110	125
Nov 1	May 1	75	80	95	110	125
Dec 1	June 1	80	85	95	115	140

remote mountaintops, or in extremely cold weather). Use a wide-angle lens if possible, push-process the film, and print on high-contrast paper or increase the contrast by some other means, such as slide duplication.

An interesting way to photograph the whole sky at once, capturing the zodiacal light, Milky Way, and other faint glows in their respective places, is to aim the camera downward at a shiny spherical surface (Fig. 4.9) – one of the wide-angle mirrors found in supermarkets would be ideal, and old, round automobile hubcaps have been used with success. The camera is, of course, reflected in the middle of the picture, as is the tripod or other structure that supports it, but if the camera is a meter or so from the reflector, its reflection is tiny. Rather than being focused on infinity, the camera lens is set to focus on a distance equal to half the radius of curvature of the reflector (reflex focusing is a great help here), and the lens is set to its widest *f*-stop. The overall effective focal length is short – in the order of 10 millimeters at the center of the field – with the result that exposures as long as two or three minutes show little trailing. The overall *f*-ratio is the same as that of the camera lens alone. Fig. 4.10 shows the results that can be obtained with this technique.

Another celestial glow worthy of attention is the *Gegenschein* ('counterglow'), a faint patch visible in the night sky in the direction exactly opposite the sun, caused by interplanetary dust particles that reflect sunlight back in the direction from which it came. Although fainter than the Milky Way, the Gegenschein is visible to the naked eye under clear, dark skies and is much brighter than some of the nebulae that amateurs routinely photograph. Although the Gegenschein has seldom been photographed, I believe there is considerable scope for observing it photographically.

To locate the Gegenschein, refer to Table 4.3 and any good star atlas. The Gegenschein is always

Fig. 4.8. *The zodiacal light and Comet Ikeya-Seki were visible to the naked eye on the morning of 26 October 1965. The photo was taken by Dennis Milon from about 2100 meters above sea level in the Catalina Mountains near Tucson, Arizona. Twenty-five seconds on Tri-X Pan using a 50-mm lens at f/2 on a Miranda camera.*

Fig. 4.9. *All-sky cameras. (a) The principle: the camera photographs the sky reflected in a spherical reflector. (b) Diagram of the all-sky camera designed and built by Chris Schur and used to take Fig. 4.10.*

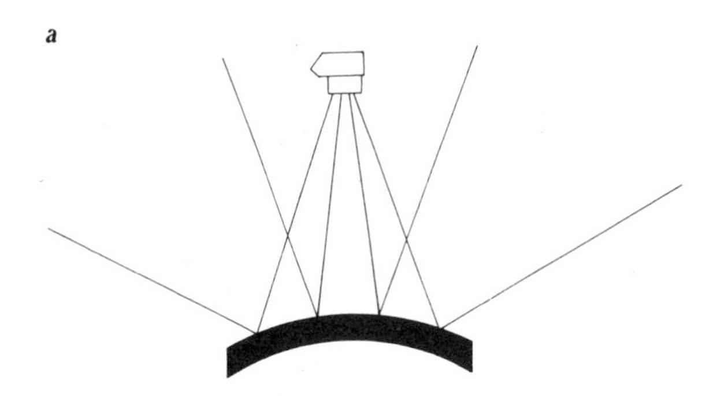

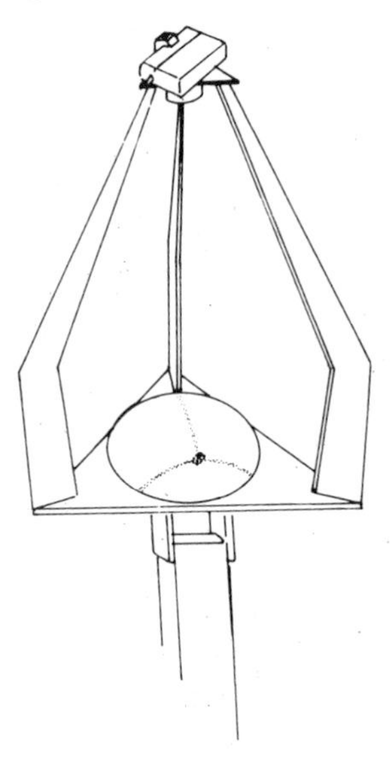

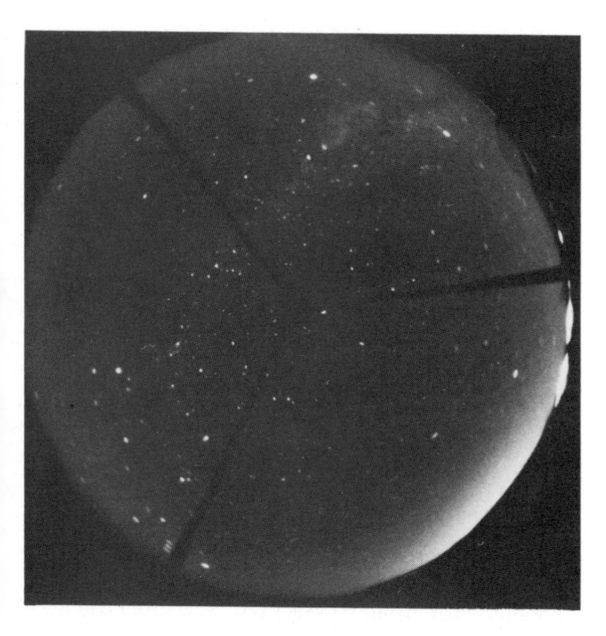

Fig. 4.10. A 10-minute exposure of the summer Milky Way on Tri-X, using the apparatus shown in Fig. 4.9, mounted on a telescope for guiding. The Great Rift is visible at the right. (Chris Schur)

Table 4.3. Where to look for the Gegenschein
For dates in between those listed, add one degree of ecliptic longitude per day.

Date	Ecliptic longitude of sun	Ecliptic longitude of Gegenschein	Constellation in which Gegenschien appears
Jan 1	280°	100°	Gemini
Feb 1	312°	132°	Cancer
March 1	340°	160°	Leo
April 1	11°	191°	Virgo
May 1	40°	220°	Libra
June 1	70°	250°	Ophiuchus*
July 1	99°	279°	Sagittarius*
Aug 1	128°	308°	Capricornus
Sept 1	158°	338°	Aquarius
Oct 1	187°	7°	Pisces
Nov 1	217°	37°	Aries
Dec 1	248°	68°	Taurus*

* The Gegenschein coincides with the Milky Way, and therefore cannot be.seen, in June, July, and December.

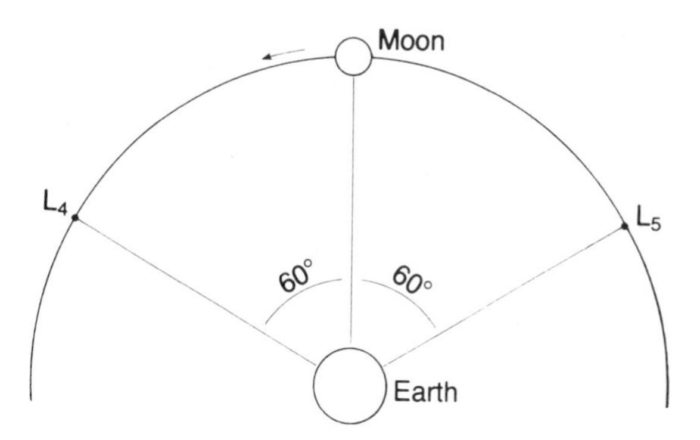

Fig. 4.11. The lunar libration clouds are swarms of dust particles that accumulate at the Lagrange points L_4 and L_5, 60° ahead of an behind the moon in its orbit.

centered on the ecliptic; the Table gives its position in terms of ecliptic longitude (the numbers marked on the ecliptic in star atlases). To photograph it, make as long an exposure as your technique permits – either a 3- to 5-minute fixed-camera star trail, or a clock-driven exposure of 20 to 30 minutes or as long as sky fog will allow. Use a normal lens, and develop for high speed and contrast. To ensure that you are not led astray by reflections or lens flare, make two exposures with the camera aimed slightly differently for each (the Gegenschein should stay in the same place relative to the stars). In any case, don't put the Gegenschein in the exact center of the field, since many lenses render the center of the field noticeably brighter than the edges whether it really is or not.

If you succeed in photographing the Gegenschein, the next things you should go for are the *lunar libration clouds* or *Kordylewski clouds,* which have the distinction of being the only objects in the sky that are regularly visible to the unaided eye but whose existence went unnoticed until the twentieth century.

The story begins in 1772, when the mathematician G. L. Lagrange was working out some of the consequences of Newton's theory of gravitation. He showed that if you have two relatively heavy objects, such as the earth and the moon or the sun and a planet, orbiting around their common center of gravity, then any small objects (such as rocks) that come into the vicinity with sufficiently low relative velocities will tend to cluster near, and wobble slowly back and forth around, the *libration points* L_4 and L_5 (Fig. 4.11). ('Libration' means 'wobble'; the term is also applied to slight variations in the moon's orientation relative to the earth.)

Early in this century, groups of asteroids (the 'Trojans') were discovered at the libration points of the sun–Jupiter system, and in the 1950s the Polish astronomer K. Kordylewski began searching for objects at the libration points of the earth–moon system. What he found was a pair of faint clouds that are comparable in brightness to the Gegenschein and are visible to the unaided eye under very dark, clear skies (most observations of them have been made from mountains, deserts or ships at sea). The existence of Kordylewski's clouds was disputed until they were observed from outer space by the Orbiting Solar Observatory satellite, OSO-6, in 1969 and 1970.

The Kordylewski clouds are 60° away from the moon in either direction along the ecliptic. It is, however, not unusual for them to be as much as 6 to 10 degrees off their predicted positions, in any direction. Each Kordylewski cloud is brightest when it is directly opposite the sun, near the Gegenschein. This means that the best time to look for the two clouds is four to six days before or after full moon respectively. The moon itself must not be in the sky at the time, of course. The Kordylewski clouds are best observed when the ecliptic is high in the sky (in the winter or spring for evening observers), but if one of them falls on the Milky Way you will not be able to see it.

Very few photographs of the Kordylewski clouds have been taken, and any that you are able to obtain will be of considerable scientific interest. In 1966, J. Wesley Simpson succeeded in capturing the Kordylewski clouds in 8-minute exposures on highly push-processed Plus-X Pan film, but the clouds were only barely visible on the negatives, and densitometer tracings were needed to confirm their presence. Nonetheless, several interesting facts emerged: the clouds wandered around quite a bit relative to their predicted positions, and the L_5 cloud turned out to be (at the time, at least) double. (See Simpson's article, 'Lunar libration-cloud photography', in *The Zodiacal Light and the Interplanetary Medium*, edited by J. L. Weinberg [Washington: NASA, 1967], pp. 97–107.)

With recent improvements in technology, there is no reason why amateurs should not expect equal or better results. The procedure is the same as for photographing the Gegenschein: go to a site well away from city lights, and make an exposure, preferably clock-driven, that is long enough to show an appreciable amount of sky fog. Then increase the contrast of the picture by printing on extra-hard paper or by repeated slide duplication, and look for irregularities or bright spots in the sky fog. As with the Gegenschein, pictures should be made in pairs in order to distinguish glows in the sky from reflections in the lens or flaws in the film.

Part II
ADVANCED TECHNIQUES

5
Optical configurations for astrophotography

Prime focus astrophotography

A telescope, like a camera lens, is basically a system for forming an image of an object; the difference between the two is that, whereas the camera lens forms its image on the film, the image formed by a telescope is fed into an eyepiece for magnified viewing.

The simplest way to take pictures through a telescope, then, is to substitute it for a camera lens. In practice this means that the camera body – normally a single-lens reflex (SLR) – is put in place of the eyepiece, as shown in the topmost of the diagrams in Fig. 5.1. Commercially made adapters for fitting cameras into telescope eyepiece tubes are readily available; Figs. 5.2 and 5.3 show the results that can be obtained.

The effective focal length of such a setup, here abbreviated F, is simply the focal length of the telescope objective:

F = focal length of telescope objective

Fig. 5.1. The five basic optical configurations for astrophotography.

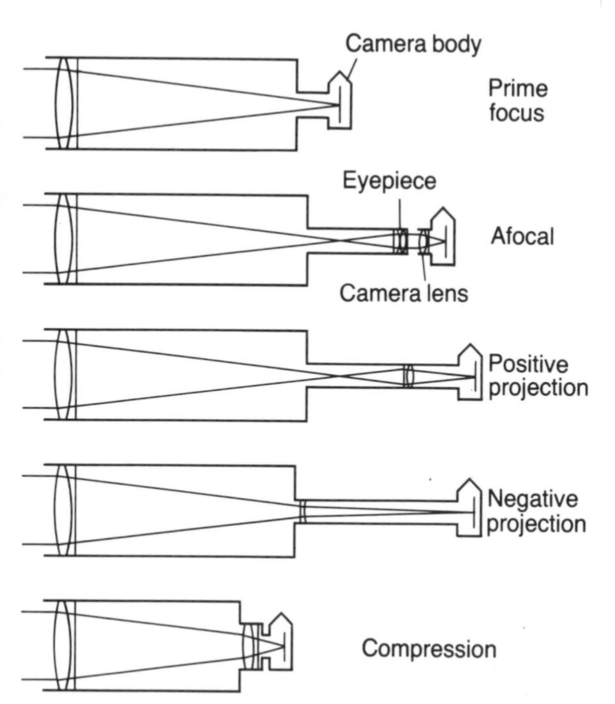

Fig. 5.2. No, this isn't the moon setting – it's Venus, photographed at the prime focus of a 32-cm (12.5-inch) telescope, in exceptionally steady air. (Jim Baumgardt)

Fig. 5.3. *This dramatic picture of Jupiter and its four Galilean satellites over the Sierra Nevada mountains was taken at the prime focus of a 20-cm (8-inch) f/6 reflector. Ordinarily, an afocal or projection setup would be used to get a larger image of the planet. (Jim Baumgardt)*

The *f*-ratio is defined as *F* divided by the telescope aperture (the diameter of the main lens or mirror):

$f = F$ / telescope aperture

In this and all calculations, the focal length and aperture must be expressed in the same units, normally millimeters. For example, a 60-mm (2.4-inch) refractor with a focal length of 900 mm operates at $f/15$ ($900 / 60 = 15$).

Fig. 5.4 shows the five most commonly encountered types of telescopes; prime focus photography is possible with all of them, though, as we shall see, some are more suitable than others. The most important constraint is the need for *back focus* (back focal distance), as illustrated in Fig. 5.5 – if you are going to put a camera body in place of the eyepiece and focus the image on the film, you must be able to rack the telescope's focusing tube inward some distance beyond the image plane. The back focus requirement of a typical 35-mm SLR is about 5 cm (2 inches), and not all telescopes will give 5 cm of back focus. With refractors and Cassegrains there is no problem, and with Schmidt–Cassegrains and Maksutov–Cassegrains the back focus is normally quite ample (as much as 40 cm or 15 inches with the Celestron 8).

The problem arises with Newtonians, which normally provide only a centimeter or two of back focus, just enough to adjust for differences between eyepieces. The back focus of a Newtonian can be increased by moving the main mirror mount about 5 cm forward from its original position in the tube. When this is done, it may be necessary to insert a small extension tube when using eyepieces, to get them out to the new position of the image plane; a suitable extension tube (for 1¼-inch or 32-mm diameter eyepieces) is easily made out of a 1¼-inch (35-mm o.d.) sink trap extension, available at any hardware store. It may also be necessary to enlarge the diagonal mirror to get full edge-of-field illumination with low-power eyepieces; alternatively, it is possible to replace the focusing mount (eyepiece holder) with a special low-profile unit that takes up less space. On the whole, however, it may be more practical to leave the telescope unmodified and use one of the other optical

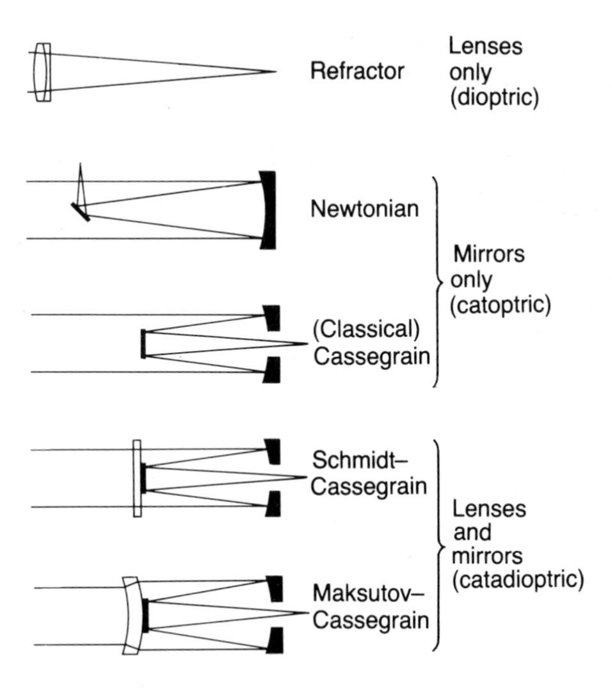

Fig. 5.4. *Five types of telescope. For convenience, the other diagrams in this chapter show the telescope as a refractor, but the principles illustrated are equally applicable to all five types.*

Fig. 5.5. *Illustrating the need for back focus.*

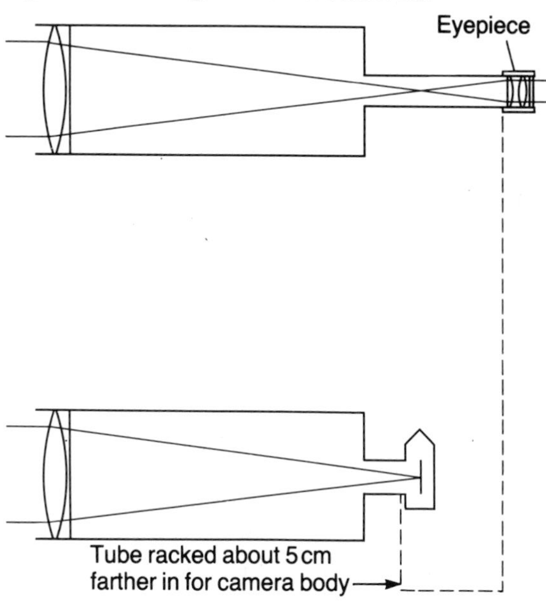

configurations that do not require as much back focus, such as afocal or positive projection.

The inside diameter of the eyepiece tube places an absolute limit on the size of the fully illuminated image area on the film; obviously, if there are no additional lenses to magnify it, the image will be no larger than the smallest opening that it has to pass through. To fill the 35-mm frame, a telescope has to have a 2-inch (50-mm) diameter eyepiece tube, together with suitably large glare stops or secondary mirror. (Some Schmidt–Cassegrains and Maksutov–Cassegrains constitute an exception; although the light forming the image has to pass through a small opening, light for different parts of the picture is traveling in different directions as it does so and thus fans out to fill the frame acceptably.) In practice, the field reduction that results from a standard-size eyepiece tube is not a serious problem; most astronomical photographs show a single object on a dark background, and those that do not can be cropped.

Fig. 5.6. *The author's 12.5-cm (5-inch) f/10 Schmidt–Cassegrain telescope set up for prime focus photography. A camera adapter fits in place of the eyepiece holder and couples to the camera by means of a T-ring. A focusing magnifier is attached to the viewfinder of the camera.*

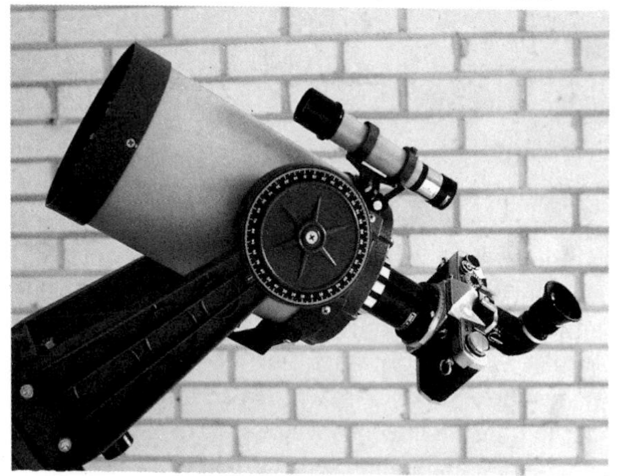

Table 5.1. *Diameter of the coma-free image area of a Newtonian or classical Cassegrain reflector*

f/	*D*	
	(mm)	(inches)
3	4.5	1/6
3.5	6	1/4
4	8	1/3
4.5	10	3/8
5	12	1/2
6	18	3/4
8	32	1 1/4
10	50	2
12	72	2 3/4
15	112	4 3/8
20	200	8

Telescope design and optical limitations

Because of the geometrical properties of light, no optical system, no matter how well manufactured, can form a perfect image of an extended object. There are six basic optical limitations on performance, or *aberrations*, that the telescope designer has to try to minimize:

Spherical aberration, which causes the periphery of the lens to have a different focal length than the center, so that rays from the same source come to different foci depending on what part of the lens they pass through.

Astigmatism, which results from the lens not being perfectly symmetrical about its center, so that out-of-focus star images are elliptical rather than round.

Coma, which blurs images in proportion to their distance from the center of the field.

Curvature of field, which causes the focal 'plane' to be curved rather than flat.

Distortion ('pincushion' or 'barrel'), which alters the shape of images that extend some distance off the axis.

Chromatic aberration, which causes light of different wavelengths (colors) to focus in different planes.

In practice, astigmatism (on axis) never occurs in well-manufactured telescopes; if you detect signs of it, check for poor collimation (misaligned optical elements), a diagonal that isn't on straight, uneven temperatures within the telescope, a defective eyepiece, or uncorrected astigmatism in your eye. Distortion and curvature of field are negligible in astronomical telescopes because of the narrow field of view. The remaining aberrations – spherical, coma, and chromatic – are dealt with differently in each of the five major telescope designs.

Chromatic aberration arises only when light is bent by passing it through a transparent medium, and the only type of telescope in which it poses a significant challenge is the refractor. The objective lens of a refractor consists of two elements, made of different types of glass, whose chromatic aberrations roughly cancel each other out; the combination is called an *achromat*. The lenses are usually figured in such a way as to correct for spherical aberration and coma as well; a well-made refractor can produce superb images over a fairly wide field, and if the *f*-ratio is about 10 or greater, the aberrations are negligible.

The problem with refractors is that an objective that is free of chromatic aberration as far as visual work is concerned may not be suitable for photography. The reason has to do with the way the chromatic aberrations of the two elements are made to cancel out: the designer chooses two wavelengths fairly far apart and causes them to focus in exactly the same plane, so that all wavelengths in between also come to more or less the same focus. For visual telescopes, the two wavelengths are normally 486 nanometers (deep blue) and 687 nanometers (deep red), at opposite ends of the visual spectrum, so that the image as seen by the eye is practically perfect.

Below 486 nm, in the violet and ultraviolet regions, chromatic aberration is still present in full force – and it is in this range that photographic films are most sensitive. The remedy is to add a yellow filter, such as a Wratten #8, #12, or #15, to eliminate deep blue, violet, and ultraviolet light; the problem then disappears, and the images look as good to the film as to the eye, though a substantial amount of light is lost. Alternatives are to use a 'photographically corrected' objective (corrected for 405–486 nm, ideal for the old blue-sensitive plates but not for modern panchromatic emulsions); a 'photovisual' objective (a compromise between photographic and visual correction); or an *apochromat*, which is corrected for three wavelengths rather than two (and costs about five times as much as an achromat). If the telescope is relatively small, none of these measures may be necessary; the images from a visual objective may be satisfactory even without a filter.

The Newtonian reflector is immune to chromatic aberration, since the light never passes through glass, and is corrected for spherical aberration by making the main mirror a paraboloid. Its main problem is coma, which can be severe at short *f*-ratios. As a rule of thumb, the diameter in millimeters, D, of the image area within which coma is negligible is half the square of the *f*-ratio:

$$D = f^2 / 2$$

This assumes a degree of resolution suitable for 35-mm photography; a larger amount of coma is tolerable in photographs on large plates that are not to be enlarged much.

As Table 5.1 shows, an $f/10$ or slower Newtonian is practically coma-free over the entire 35-mm field; an $f/8$ is coma-free over as much of the image as can get through a standard-size eyepiece tube; but $f/6$ and faster telescopes are coma-free only over small areas. This explains why ultra-fast Newtonians are suitable only for low-power observing or deep-sky photography in which maximum resolution of fine detail is not necessary. It also underscores the need for precise collimation: if what looks like the center of the field of an $f/4$ telescope is actually off axis by a mere 12 millimeters (half inch), coma will be severe. (Incidentally, since the linear size of the coma-free field depends only on the *f*-ratio and not on the focal length, the 5-meter, 200-inch, $f/3.3$ telescope on Mount Palomar should theoretically give good images only within about 3 millimeters of the axis. In practice, special corrector lenses that counteract coma are used near the film.)

The classical Cassegrain system consists of a paraboloidal primary mirror plus a hyperboloidal secondary that multiplies the effective focal length by a factor of four or five. The resulting system is very compact in proportion to its focal length; a telescope with a focal length of 5 meters (15 ft) may be physically only 1 meter (3 ft) long. Moreover, since the primary mirror is a paraboloid it can be used equally well as the mirror of a Newtonian; some telescopes are constructed so that they can be used, for example, either as $f/4$ Newtonians, or as $f/15$ Cassegrains. The off-axis coma of a classical Cassegrain is the same as that of a Newtonian of the same *f*-ratio, but Cassegrains usually operate at $f/15$ or $f/20$, so coma is rarely a problem. The classical Cassegrain is ideal for lunar and planetary work.

The Schmidt–Cassegrain and Maksutov–Cassegrain represent modifications of the Cassegrain design that reduce off-axis aberrations almost to zero and make it possible to fill an exceptionally large field with sharp images. In both cases the primary mirror is spherical, and spherical aberration is dealt with by inserting a glass corrector plate to cancel it out. The Schmidt–Cassegrain corrector plate is nearly flat (it is thickest in the center, gets thinner as you move out to just over half its radius, and then gets thicker again towards the edge); the Maksutov corrector has two spherical surfaces of nearly the same curvature. The chromatic aberration introduced by the corrector plate is negligible in either case. The Maksutov

(a) The angular diameter of the moon as seen from earth is one half degree.

(b) The size of the image on the film depends on the angular diameter and the focal length.

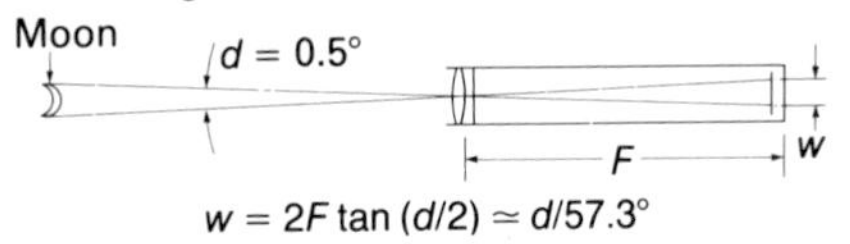

Fig. 5.7. *Angular diameter and image size.*

design is more common in very small telescopes, such as the 9-cm (3½-inch) Questar and Celestron C90, and in the 'mirror lenses' offered by camera makers such as Nikon and Minolta.

The Schmidt–Cassegrain and Maksutov–Cassegrain are in many ways the best all-round astrophotographic instruments: they offer freedom from aberrations over a wide field with generous amounts of back focus, and their compact design (usually incorporating a fork-type equatorial mount with manual slow motions and a built-in drive) makes them convenient to use. Their main disadvantage is their speed, typically only $f/10$.

Image size and field of view

Because the sky appears to be an infinite distance away, the apparent sizes of celestial objects are expressed as angles rather than linear distances; the observer's eye, or the center of the telescope objective, is the vertex of the angle, as shown in Fig. 5.7. The units used for measuring angles are the following:

1 degree (1°) = 60 arc-minutes (60′)

1 arc-minute (1′) = 60 arc-seconds (60″)

1 radian = 57.3 degrees = 3438 arc-minutes = 206 265 arc-seconds

The lower part of Fig. 5.7 shows what is involved in calculating the width, w, of the image that an object whose apparent diameter is d will form on the film when the effective focal length is F. The precise formula, derived from basic trigonometry, is:

$$w = 2\,F \tan(d/2)$$

Focal length (mm)	Field of view (on 35-mm film	Moon image (mm)	Jupiter image on film (mm)	Jupiter image on ×15 enlargement (mm)
400	3.4° × 5.2°	3.6	–	–
500	2.7° × 4.1°	4.5	–	–
600	2.3° × 3.4°	5.4	–	–
700	2.0° × 2.9°	6.4	–	–
800	1.7° × 2.6°	7.3	–	–
1000	1.4° × 2.1°	9.1	0.2	3.0
1250	1.1° × 1.7°	11	0.25	3.8
1500	0.9° × 1.4°	14	0.3	4.5
2000	41′ × 62′	18	0.4	6.0
2500	33′ × 50′	23	0.5	7.5
3000	28′ × 41′	27	0.6	9.0
4000	21′ × 31′	36	0.8	12
5000	17′ × 25′	45	1.0	15
6000	14′ × 21′	55	1.2	18
8000	10′ × 15′	73	1.6	24
10 000	8.3′ × 12′	91	2.0	30
12 000	6.9′ × 10′	109	2.4	36
14 000	5.9′ × 8.8′	127	2.8	42
16 000	5.2′ × 7.7′	145	3.2	48
18 000	4.6′ × 6.9′	164	3.6	54
20 000	4.1′ × 6.2′	182	4.0	60

Table 5.2. *Field of view, moon image size, and Jupiter image size for a variety of focal lengths*

But the formula can be simplified by noting that, with the small angles involved in astrophotography, the tangent of any angle d is very close to the value of d itself in radians; the difference is less than 1% for angles as large as 10°. The simpler formulae are as follows:

$w = F \times d$ (d in radians)

$w = (F \times d)\,/\,57.3$ (d in degrees)

$w = (F \times d)\,/\,3438$ (d in arc-minutes)

$w = (F \times d)\,/\,206\,265$ (d in arc-seconds)

(F and w are always given in the same units, usually millimeters.) The apparent diameter of the moon is just over 30 arc-minutes, and that of Jupiter averages about 40 arc-seconds, leading to two easy-to-remember rules of thumb:

moon image size $= F / 110$

Jupiter image size $= F / 5000$

where F and the image size are expressed in the same units. It is not difficult to remember the size of other objects by comparing them to the moon or Jupiter.

But what do you do if you know the image scale and you want to know the focal length? There are two situations in which such a question can come up: deciding what focal length to use to photograph a particular object, or calculating the effective focal length of your telescope from the size of the image of an object of known angular dimensions. (The latter is common with projection setups where the lens positions are not known precisely.) The exact formula in such a case is:

$$F = w / (2 \tan (d/2))$$

and, just as before, it can be simplified:

$F = w / d$ (d in radians)

$F = 57.3 \times (w / d)$ (d in degrees)

$F = 3438 \times (w / d)$ (d in arc-minutes)

$F = 206\,265 \times (w / d)$ (d in arc-seconds)

The exact angular sizes of the moon and planets on a given date can be obtained from the *Astronomical Almanac*. Brief exposures of bright double stars of known separation are also useful for measuring focal length; for data on particular stars, see Robert Burnham, Jr., *Burnham's Celestial Handbook* (3 vols., New York: Dover, 1978–9).

Finally, there is the question of how wide an angular field the picture will cover (assuming the eyepiece tube leaves the film fully illuminated). This is equivalent to asking how big an object has to be – in angular terms – in order for its image to stretch from one edge of the film to the other; we therefore set w equal to the width of the picture area on the film and solve for d. The precise formula is:

$$d = 2 \arctan (w / (2F))$$

The simplified versions, which are accurate to within 1% when F is at least 100 mm, are:

d (in radians) $= w / F$

d (in degrees) $= 57.3 \times (w / F)$

d (in arc-minutes) $= 3438 \times (w / F)$

d (in arc-seconds) $= 206\,265 \times (w / F)$

If the film format is not square, there will of course be two values for w, one for the long dimension and one for the short dimension. The dimensions of some commonly used film formats are:

120 film (6-cm square format): 55 mm square

120 film (4.5 × 6-cm format): 42 × 55 mm

35-mm full frame: 24 × 36 mm

35-mm half frame: 18 × 24 mm

110 film: 12 × 17 mm

Table 5.2 summarizes the field of view and image sizes for the moon and Jupiter, the latter both on the film and on a ×15 enlargement, for a variety of effective focal lengths ranging from those of telephoto lenses to those used in high-resolution planetary photography. The same values apply to any given effective focal length whether it is obtained at the prime focus or by positive projection, negative projection, or any other method.

The afocal method

Prime focus photography has two major disadvantages: it requires a substantial amount of back focus, more than many telescopes can provide, and, for any given telescope, it allows photography at only one effective focal length, almost always too short to do justice to the telescope's full ability to record lunar and planetary detail. The afocal method, already introduced in Chapter 2, lacks these disadvantages and has several additional points in its favor:

1 The camera need not have through-the-lens (reflex) focusing. (The hand-held telescope method of focusing is described in Chapter 2.)

2 The eyepiece and camera lens are working at the optical image distances for which they were designed to give best performance.

3 The camera and telescope can stand on separate tripods, preventing any transfer of vibration from the camera shutter to the telescope.

Using two tripods can admittedly be awkward, particularly when a clock drive is moving the telescope relative to the camera, and it is impractical for exposures of more than a second or two. The alternative is to couple the camera to the telescope by means of some kind of bracket; suitable brackets are sometimes available commercially, and a number of designs for home-built ones are given in Sam Brown's *All About Telescopes* (Barrington, New Jersey: Edmund Scientific Company, 1967, pp. 60–1). A custom-built coupling can also solve the vexatious problem of how to get the camera lens centered relative to the eyepiece.

The basic formulae for afocal photography are as follows. The overall effective focal length, F, is given by:

$$F = \text{focal length of camera lens} \times \text{magnification of telescope}$$

or, alternatively, by a formula also applicable to positive and negative projection systems:

$$F = F_1 \times M$$

where F_1 is the focal length of the telescope objective by itself and M, the projection magnification, is given (for afocal setups) by the formula:

$$M = \text{camera lens focal length} \,/\, \text{eyepiece focal length}$$

The f-ratio is given by:

$$f = F / D = f_1 \times M$$

where D is the diameter of the telescope objective and f_1 is the f-ratio of the telescope objective by itself.

All types of eyepieces can be used in afocal photography, but those that give good eye relief (orthoscopics, RKEs, symmetricals, or Plössls) are best. Ideally, the diaphragm of the camera lens should be located at the exit pupil, but in practice the construction of the lens usually makes this impossible, and the lens is (from the theoretical point of view) always too far from the eyepiece and, therefore, sees only the middle of the field. (With some eyepieces that give exceptional amounts of eye relief, such as the 28-mm RKE, it is possible to put the camera too close to the eyepiece; experiment with the telescope aimed at a terrestrial object to find which camera position gives the least vignetting.)

The camera is normally fitted with a normal or medium telephoto lens; nothing is gained by using a wide-angle, since the eyepiece limits the field of view. In theory, the camera lens should be set at the f-stop equal to the f-ratio of the overall system, but in practice it is usually left wide open to allow for errors of centering.

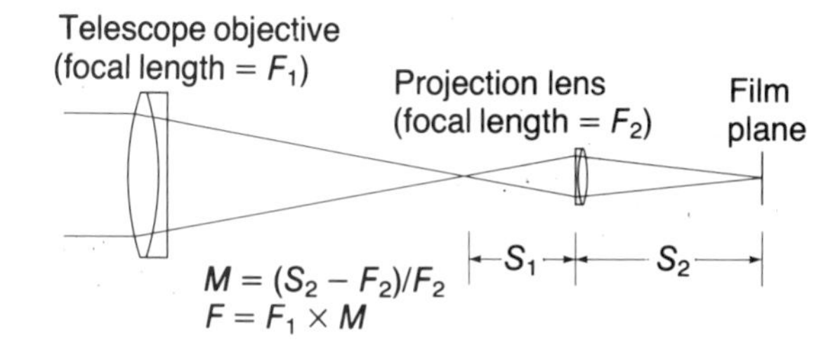

Fig. 5.8. Positive projection.

Note that M can be less than 1; that is, you can use the afocal method to decrease rather than increase the system focal length. This can be helpful in photographing faint objects, because it reduces the system *f*-ratio and hence the exposure time. An afocal system with $M = 0.17$ has been used with the 2-meter (82-inch) $f/12$ Cassegrain at McDonald Observatory to yield effective $f/2$; it consists of an ordinary Leitz Summicron camera lens fed by a specially constructed giant eyepiece (see A. B. Meinel, 'An $f/2$ Cassegrain camera', *Astrophysical Journal*, vol. 124, pp. 652–4, 1956). With amateur telescopes it is rarely practical to take M below about 0.5; however, since the exposure time necessary to record a given object varies as the square of f (and hence as the square of M), even a configuration with $M = 0.7$ cuts the exposure time in half as compared to prime focus photography.

Remember that the camera lens can accommodate a larger exit pupil, and hence a longer focal-length eyepiece with a given telescope, than can the eye; you may want to experiment with building giant eyepieces. Good results have been reported with a setup in which a single achromatic lens (a 7 × 35 binocular objective with the flatter side toward the camera) functions as a 176-mm eyepiece; two such lenses, placed with their more sharply curved sides toward each other and separated by about 10 mm, would make a reasonable 90-mm symmetrical eyepiece. Giant eyepieces can give prodigious amounts of eye relief; experiment to find the best camera position.

Positive projection

Positive projection involves using a small positive (convex) lens – often an eyepiece, hence the alternative term *eyepiece projection* – to form an image of an image. That is, the image formed by the telescope at its prime focus serves as the object for the projection lens, which forms an enlarged image of it on the film.

Fig. 5.8 illustrates the important parameters. The projection magnification is equal to the ratio of the lens-to-film distance (S_2 in the diagram) and the distance from the original image to the lens (S_1):

$$M = S_2 / S_1$$

In practice, however, S_1 is usually not known; the known parameters are S_2 and the focal length of the projection lens, F_2, and the relevant formula is:

$$M = (S_2 - F_2) / F_2$$

or, if you want to find the appropriate lens-to-film distance to get a particular magnification:

$$S_2 = F_2 \times (M + 1)$$

Given M, the effective focal length and *f*-ratio of the whole system can then be calculated from the same formulae we used on the facing page for the afocal method:

$$F = F_1 \times M$$
$$f = F / D = f_1 \times M$$

Finally, S_1, should you ever need it, is given by:

$$S_1 = S_2 / M$$

Notice that, if the magnification is to be greater than 1, S_2 must be greater than $2 \times F_2$. Magnifications of less than 1 are, of course, quite possible, and can be used to reduce the *f*-ratio; they occur when the value of S_2 is less than $2 \times F_2$ but greater than F_2. If S_2 is less than F_2, no image forms at all. Likewise, S_1 has to be greater than F_2 (it would be equal to F_2 if the positive lens were being used as an eyepiece rather than a projection lens), and hence an eyepiece being used for projection has to be racked *outward* from its usual position – positive projection requires, so to speak, a negative amount of back focus compared to an eyepiece.

Let's take a concrete example. Suppose we are using a 150-mm (6-inch) reflector of 1200 mm focal length and the projection lens is an 18-mm eyepiece placed 75 mm (3 inches) from the film, and we want to know the projection magnification:

$$M = (S_2 - F_2) / F_2 = (75 - 18) / 18 = 57/18 = 3.167$$

That is, the projected image is a bit more than three times the size of the image at the prime focus. Given M, we find the overall effective focal length and f-ratio as follows:

$$F = F_1 \times M = 1200 \times 3.167 = 3800 \text{ mm}$$

$$f = f_1 \times M = 8 \times 3.167 = 25.3$$

That is, the overall system has an effective focal length of 3800 mm and operates at f/25.3.

Eyepieces are the most commonly used projection lenses, but not necessarily the best ones. Projection is not really what they are designed for, and many astrophotographers have obtained better results by using enlarging lenses, movie or subminiature camera lenses, or microscope objectives, all of which are more highly corrected than eyepieces for the conditions encountered in projection astrophotography. A camera or enlarger lens used in such a setup should be mounted backward, so that the side which faced away from the film in the original camera or enlarger now faces toward it. The reason is that, although from the standpoint of the lens design it does not matter whether the light rays are passing through the lens from front to back or back to front, it does matter a great deal which of the two object/image distances – S_1 or S_2 – is the larger. A camera lens normally focuses an image of a distant object onto a nearby piece of film; when the object (that is, the original telescope image which is to be projected) is closer to the lens than to the film, the lens should be mounted backward. (For the same reason, lenses are reversed in extreme close-up photography.)

Naturally, the camera's normal lens can be used as a projection lens, though at higher magnifications, S_1 can become awkwardly large (over 30 cm or a foot for ×6 magnification with a 50-mm lens); a wide-angle lens or, better, the normal lens of a smaller-format camera can be easier to use. A cheap way to get good lenses is to salvage them from junked movie cameras. (Lenses from projectors cannot, however, be relied upon; they are generally designed for brightness rather than sharpness.)

Microscope objectives are particularly suitable for projection systems because positive projection so closely resembles the application for which they were originally designed. A microscope objective yields its rated magnification when S_2 is about 160 mm; more generally, the focal length of a microscope objective is roughly equal to 160 mm divided by the rated magnification plus one (that is, 14.5 mm, or 160/(10 + 1), for a ×10 objective), and S_2 can be varied to get different magnifications. Given a choice, select microscope objectives designed for use on specimens that are not under a cover glass, as in watchmaking or mineralogy; conventional biomedical-type objectives also work well, but oil-immersion lenses are obviously unsuitable.

How large should the projection lens be in order to catch all of the relevant light rays? This question cannot be answered precisely without ray-tracing the whole optical system, but the following formula can serve as a rule of thumb:

$$\text{minimum diameter} = ((M + 1) / M) \times (F_2 / f_1)$$

(Here f_1 is the f-ratio of the telescope objective.) For example, in the setup discussed a few paragraphs back, the clear aperture of the field lens of the 18-mm eyepiece should be at least:

$$((3.167 + 1) / 3.167) \times (18 / 8) = 1.32 \times 2.25 = 2.96 \text{ mm}$$

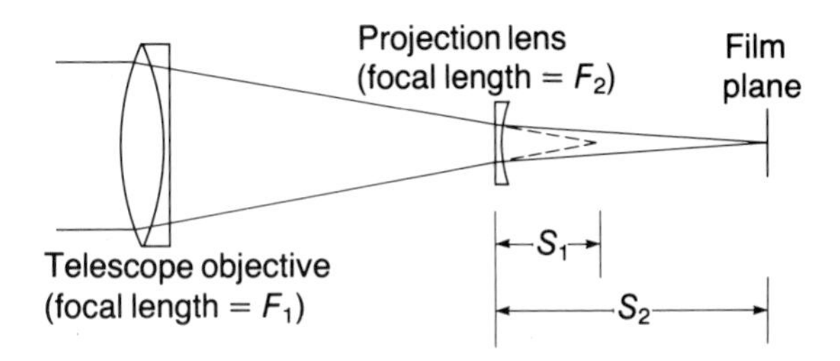

Fig. 5.9. *Negative projection.*

This is only a rough answer, but a quick glance shows that the field lens of an 18-mm eyepiece is a good bit more than 3 millimeters across, so there is nothing to worry about.

The question is of greater interest when the projection lens has an adjustable diaphragm. The minimum aperture of the projection lens, expressed in terms of *f*-stops, is given by the formula:

$f_2 = f_1\ (M\ /\ (M + 1)\ )$

That is, the projection lens should always be faster than the main telescope objective by at least a factor of $M\ /\ (M + 1)$. Suppose that in the example above we have a movie-camera lens of 18-mm focal length instead of an 18-mm eyepiece; the smallest aperture to which we can set it with impunity is:

$8 \times (3.167\ /\ (3.167 + 1)\ ) = 8 \times 0.76 = f/6.1$

Finally, the numerical aperture (NA) of a microscope objective is defined in terms of S_1 rather than focal length, leading to a much simpler formula:

minimum NA $= 0.5\ /\ f_1$

For example, an $f/8$ telescope requires a microscope objective with a numerical aperture of 0.0625 or greater. Since only a few very low-powered microscope objectives have NA less than 0.10, problems are unlikely to arise.

Negative projection

Another popular setup involves placing a negative (concave) lens in the path of the light coming from the telescope; its effect is to make the image larger and the effective focal length longer (Fig. 5.9). The formulae for negative projection are the same as those for positive projection (repeated here for convenience) so long as you remember that *the focal length of a negative lens is a negative number.*

$M = S_2\ /\ S_1 = (S_2 - F_2)\ /\ F_2$

$F = F_1 \times M$

$f = F\ /\ D = f_1 \times M$

$S_1 = S_2\ /\ M$

To take a concrete example, suppose that the telescope is a 150-mm (6-inch) $f/8$ reflector of 1200-mm focal length (as in the previous example), and the projection lens, of −45 mm focal length, is located 75 mm from the film. Then:

$M = (75 - (-45)\ )\ /\ (-45) = 120/(-45) = -2.67.$

M comes out negative because the magnified image is on the same side of the lens as the original image; its negative sign should be ignored in the subsequent formulae, giving:

$F = 1200 \times 2.67 = 3200$ mm, and
$f = 8 \times 2.67 = 21.3.$

Moreover, S_1 also comes out negative – again, because it is measured on the same side of the lens as S_2, rather than on the opposite side – and in the present example is:

$S_1 = S_2\ /\ M = 75/(-2.67) = -28.1$ mm.

With negative projection setups, the value of S_1 is always between zero and F_2, and magnifications of less than unity are not possible. Ignoring the negative sign, S_1 of course represents the amount of back focus required by the setup – never zero, but often small enough to make it practical to use an unmodified Newtonian telescope, particularly if the projection lens can be placed inside the eyepiece tube.

Two types of lenses are commonly used for negative projection: Barlow lenses originally designed for visual use with telescopes, and teleconverters intended for use with telephoto lenses on cameras. In my experience, teleconverters are the

5.10

5.11

Fig. 5.10. *A negative projection setup consisting of a 12.5-cm (5-inch) f/10 Schmidt–Cassegrain telescope, a ×3 teleconverter, a camera body, and a focusing magnifier.*

Fig. 5.11. *The crescent moon photographed with the setup in Fig. 5.10. Agfa Superpan (ISO 200), 1/4 second.*

better of the two; besides being slightly less expensive, they give good images over a wide, flat field, and they often perform even better with telescopes than with the telephoto lenses for which they were originally designed. If the teleconverter is mounted on the camera in the normal way, M is equal to the rated magnification taken as a negative number (that is, with a ×2 converter, $M = -2$), and the back focus requirement of the combination is the same as that of the camera body by itself. The magnification can be increased by putting an extension tube between the converter and the camera.

There are several ways to find the focal length of a negative projection lens. To begin with, if you know the rated magnification and the lens-to-film distance S_2 (or, in the case of a Barlow designed for visual work, the distance from the lens to the field stop of the eyepiece), you can use the formula:

$$F_2 = S_2 / (M + 1)$$

(again, remember that M is negative). Alternatively, you can determine the focal length of any negative lens with only a ruler and a magnifying glass, using the following procedure:

1 Find the focal length of the magnifying glass by using it to form an image of the sun (or some other distant object) on a piece of paper and measuring the distance from the lens to the paper. Call this distance F_P.

2 Place the negative lens and the magnifying glass in contact with each other to find out which of the two is stronger. If the combination acts as a magnifying glass, go to step 3; if it acts as a reducing glass, go to step 4.

3 Determine the focal length of the combination using the same technique as in step 1; call it F_C (a positive number). Then apply the formula:

$$\text{focal length of negative lens} = (F_P \times F_C) / (F_P - F_C)$$

Here F_P is negative, and hence the focal length of the negative lens is negative.

4 Use the negative lens and the magnifying glass to make a Galilean telescope. That is, hold the negative lens to your eye, hold the positive lens next to it, and slowly move the positive lens away until you get a clear, magnified view of distant objects. Measure the separation of the two lenses when the image is in focus; call it S. Then:

$$\text{focal length of negative lens} = S - F_P.$$

S is always less than F_P.

A negative achromat to be used as a projection lens should be placed with the flatter side away from the film to minimize aberrations.

Compression

Compression (Fig. 5.12) is the opposite of negative projection: the lens inserted into the converging light cone is a positive lens and serves to make the image smaller rather than larger. (The rationale for doing this is to reduce the *f*-ratio and hence the exposure time needed for photographing faint objects.) The same formulae apply; note that although F_2 is positive, S_1 and M are still negative, just as in negative projection.

The compressor lens (also referred to as a telecompressor or rich-field adapter) is normally a simple achromat – often a salvaged binocular objective – mounted with its flatter side toward the film. The lens should be as large in diameter as the eyepiece tube through which the light reaches it; even so, some vignetting is inevitable, since the size of the overall image is being reduced. Off-axis aberrations are also likely to be noticeable; in fact, they are practically the trademark of photographs made by this method.

Fig. 5.12. Compression.

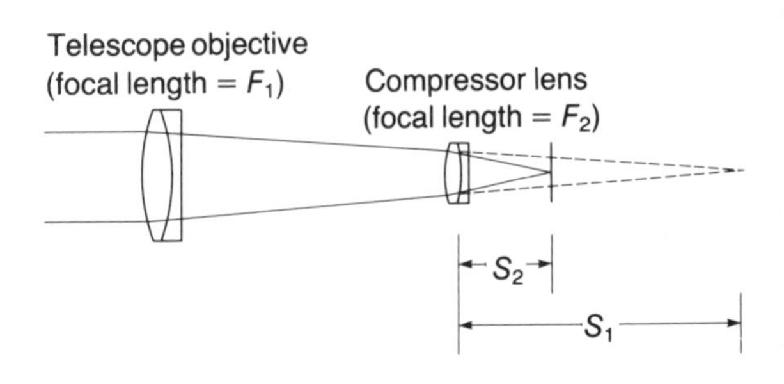

As an example, consider a 125-mm (5-inch) Schmidt–Cassegrain of 1250 mm focal length, used with a compressor lens of 200 mm focal length placed 75 mm from the film. The calculations are as follows:

$$M = (S_2 - F_2) / F_2 = (75 - 200) / 200 = (-125) / 200 = -0.625$$

$$F = F_1 \times M = 1250 \times 0.625 = 781 \text{ mm}$$

$$f = f_1 \times M = 10 \times 0.625 = 6.25$$

(In computing F and f, the negative sign of M is ignored.) The compressor lens effectively transforms an $f/10$ telescope into an $f/6.25$ system that requires exposure times only 40% as long as the original.

The catch is that the amount of back focus required is rather large. In the example we are considering:

$$S_1 = S_2 / M = 75 / (-0.625) = -120 \text{ mm}$$

That is, the setup requires 120 mm (nearly 5 inches) of back focus even though the compressor lens is only 75 mm (3 inches) from the film.

With a compressor lens, M is always between 0 and -1; best results are usually obtained at about -0.7, which is the value needed to cut exposure times in half (exposure time is proportional to the square of M, and -0.7 squared is 0.49). The problem of lens aberrations in compression setups has not yet been solved satisfactorily; it is possible that in the near future, if the popularity of the technique continues to increase, we may see lenses available that are specially corrected for compression, instead of the simple achromats presently on the market. In the meantime, it is a good idea to remember that reduced-size images can also be obtained with the afocal method and with positive projection.

Diffraction-limited resolution

If light always traveled in straight lines, as the idealizations that underlie geometrical optics assume it does, the resolving power of a telescope would be limited only by uncorrected aberrations and manufacturing tolerances. In reality, however, light – like waves in a pond – shows a detectable tendency to bend around corners. This bending is called *diffraction* and is noticeable only with very narrow apertures or very high magnifications. The larger the telescope, the less diffraction affects the quality of the image; this is why larger telescopes show more fine detail, at a given magnification, than do smaller telescopes.

The resolving power of a telescope is traditionally equated with the *Rayleigh limit*, that is, the angular separation of two stars whose images are seen as just touching, given by the formula:

Rayleigh limit (in arc-seconds) =
0.252 × (wavelength in nm / aperture in mm)

The human eye is most sensitive at 550 nanometers; most photographic films, at about 400 nm (allowing for the ultraviolet-absorbing properties of glass). Plugging these values into the formula, we get:

Rayleigh limit (visual, in arc-seconds) =
140 / aperture in mm

Rayleigh limit (photographic) =
100 / aperture in mm

That is, a 100-mm (4-inch) telescope should in theory be able to resolve a 1.4″ double star visually and a 1.0″ double star photographically, under ideal conditions. (The *Dawes limit*, determined empirically from visual double-star work, is 115 divided by the aperture in millimeters; it is equivalent to the Rayleigh limit for a wavelength of 455 nm.)

The Rayleigh limit can be expressed in terms of

resolution of a certain number of lines per millimeter on the film; as such, it becomes a function of the *f*-ratio rather than the diameter per se. The formula is:

Rayleigh limit (in lines/mm) =
820 000 / (wavelength in nm × *f*-ratio)

or, at 400 nanometers:

Rayleigh limit (in lines/mm) = 2000 / *f*

The trouble with the Rayleigh limit is that it is defined in terms of star images *just touching* and hence applies only to extremely high-contrast objects, where the images are easy to separate from the background. Low-contrast lunar and planetary detail is generally not visible unless the images of the points in question are *cleanly separate*, and the practical limit is therefore much lower:

practical limit (in lines/mm) = 1000 / *f*

Table 5.3 summarizes the results of applying these two formulae.

Table 5.3. *Theoretical and practical resolution limits (in lines per millimeter, for various f-ratios,* at 400 nm)

f-ratio	Rayleigh limit	Practical limit
f/4	500	250*
f/8	250	125*
f/16	125	63
f/32	63	31
f/64	31	16
f/100	20	10
f/150	13	7
f/200	10	5

* *Note:* Lenses rarely resolve better than 60 or 80 lines/mm because of aberrations and manufacturing tolerances.

Now comes the practical question: how much projection magnification is enough? That is, at what *f*-ratio should a telescope system be operated in order to take full advantage of its ability to record fine detail? The following considerations apply:

1 The higher the *f*-ratio, the fainter the image and the longer the exposure time; hence *f* should be no higher than necessary.

2 A 35-mm slide looks sharp when it resolves 40 lines per millimeter; a negative that resolves 40 lines per millimeter will produce sharp-looking ×8 enlargements. In practice, slides or negatives that resolve only 20 lines per millimeter look only slightly fuzzy.

3 To bring out all the detail in a photograph, enlarge it until it ceases to look sharp – but no further. When we say a picture looks sharp, we mean it contains more detail than the viewer's eye can see. Enlarged further, to bring all the detail clearly into view, it would no longer look perfectly sharp.

If the *f*-ratio of the telescope system produces a resolution of about 10 lines per millimeter on the film, then all the detail captured by the telescope will be visible on ordinary enlargements. Moreover, the resolving power of the film will not be a limiting factor, since the low-contrast resolving power of Ektachrome 200 (for example) is about 50 lines/mm.

An *f*-ratio of about 100 gives this amount of resolution. Hence lunar and planetary photographers are quite correct in working at about *f*/100. With films of exceptional resolving power, such as Kodak Technical Pan, the ideal *f*-ratio is more like 50.

6
The solar system

High-resolution astrophotography

In photographing the sun, moon and planets, you are normally using your telescope for its ability to magnify fine detail, rather than to collect faint light. That is, your main technical goal is high resolution, not light grasp. In order to obtain maximum resolution, you need:

1 perfect optics;

2 film of adequate resolving power;

3 accurating focusing;

4 perfect tracking (or an exposure short enough that a clock drive is not necessary);

5 freedom from vibration; and

6 steady air, both inside and outside the telescope.

With modern, commercially built telescopes, the quality of the optics is rarely, if ever, a problem (except for a few cheap eyepieces); if you build your own telescope, its quality is, of course, up to you. We dealt with the resolving power of the film briefly at the end of the previous chapter; the solution is to magnify the image enough that the resolution of the film is no longer the limiting factor, which means working at or above approximately *f*/100 (Ektachrome) or *f*/50 (Technical Pan). We'll deal with focusing in the next section. That leaves perfect tracking, freedom from vibration, and steady air.

In lunar and planetary work, tracking is not quite as critical as it is in deep-sky photography because lunar and planetary exposures are shorter. It is not usually necessary (or even possible) to make guiding corrections during the exposure, nor does polar alignment have to be more accurate than to within a degree or two. You can even get by without a clock drive if you keep the exposures short enough.

If the total tracking error during the exposure is less than 0.5″, it will have no visible effect on your photographs regardless of the size of your telescope – no telescope can show detail finer than 0.5″ through the earth's turbulent atmosphere. This means that if the exposure is 1/30 second or shorter, a clock drive is unnecessary. (Relative to a fixed telescope, the sky moves at the rate of 15 arc-seconds in one second of time, or 0.5″ in 1/30 second.) Don't hesitate to try considerably longer exposure without a clock drive, up to 1/4 second or so if necessary; the resulting blur may not be obtrusive. At short focal lengths, of course, you can tolerate considerably more than 0.5″ of trailing; the amount depends on the focal length, as discussed in Chapters 1 and 2.

If you are using a clock drive, practically the only thing you have to worry about is making sure it really is tracking at the moment of exposure. Balance the telescope carefully, and after making any change in its aiming, however slight, let the drive run for a few seconds to take up slack in its gears before you make the exposure. The ordinary solar drive rate (one revolution in exactly 24 hours) is satisfactory for planetary work; there is no single 'planetary rate', since planets deviate from the solar rate by different amounts at different points in their orbits, and the deviations would in any case be noticeable only in an exposure of several hours. (The moon does move at a rate of its own; the problem of tracking it will be dealt with below.)

The telescope must not, of course, vibrate during the exposure. Many telescopes, especially small refractors, come from the factory with seriously inadequate tripods, which can be improved by bracing. If there is play in any of the joints at the top of the tripod, tighten or shim them somehow, and if the three legs are not connected at the bottom, join them with chains or rigid braces as close to the ground as possible. The idea is to make a structure consisting entirely of triangles, since a triangle is unbendable as long as the lengths of its sides remain fixed. If the tripod is made of wood, so much the better: although wood may be more flexible than

Fig. 6.1. *The author's wooden tripod, which is considerably steadier than metal tripods of similar bulk. Note that the structure consists entirely of triangles.*

metal, the crucial difference is that metal responds to a blow by 'ringing' and vibrating for some time, whereas vibrations in wood damp out quickly. (See John J. Brooks, 'Structural considerations for telescope makers', *Sky and Telescope*, June 1976, pp. 423–8. Fig. 6.1 shows the kind of tripod Brooks advocates.)

Any telescope mounting that is light enough to be portable, and many that are not, will be vulnerable to vibration caused by the camera shutter. To determine how serious this problem is with your equipment, train the telescope on a star or distant terrestrial object and look through the viewfinder while tripping the shutter. (If the camera is not an SLR, clamp it to the telescope near its usual position but look through the telescope eyepiece.) If the image shakes at all, you have a vibration problem.

To minimize camera-shutter vibration in an SLR, lock up the camera's mirror, if possible, before tripping the shutter. It goes without saying that you should use a cable release, preferably the air-bulb type, to keep from shaking the camera; better yet, use the self-timer to trip the shutter after all vibration caused by touching the equipment has died away.

Alternatively, you can eliminate shutter-induced vibration by controlling the exposure, not with the shutter, but with a black card held in front of the telescope (not touching it). Open the camera shutter, wait a couple of seconds for the vibration to die down, and then remove the card; bring the card back into position before closing the camera shutter. The shortest exposure you can get by this method is about 1/4 second. A large, dark hat is sometimes used instead of a card; hence the technique is sometimes referred to as the 'hat trick'.

The only solution to vibration caused by wind is to surround the telescope with a low wall or even a domed observatory. If you require portability, some ingenuity may be needed to devise a movable enclosure. Fortunately, strong winds in the middle of the night are uncommon in most climates; when they do occur, the air is usually so turbulent that you would get poor results even with a perfectly steady telescope.

Even when the air does not seem to be moving, it is seldom optically steady; irregularities in the air, analogous to the 'heat waves' you see in the air above a hot iron, are in fact the main obstacle to high-resolution photography. After all, distortion amounting to only half an arc-second – less than one-hundredth of the smallest amount the unaided eye can see – can keep a 12.5-cm (5-inch) telescope from reaching full resolving power.

Ordinarily, the blur or wobble is more like 2 or 3 arc-seconds; if it gets down to 1 arc-second, conditions are distinctly better than average. You can estimate the amount of atmospheric blurring by observing a close double star of known separation. There are two different effects to contend with: a continuous blurring that makes the image look out of focus and clears up only for brief moments, and a slow unsteadiness that makes the whole image move around a bit. Both of these interfere with photography, though the first is the only one a visual observer is likely to notice (if the whole image

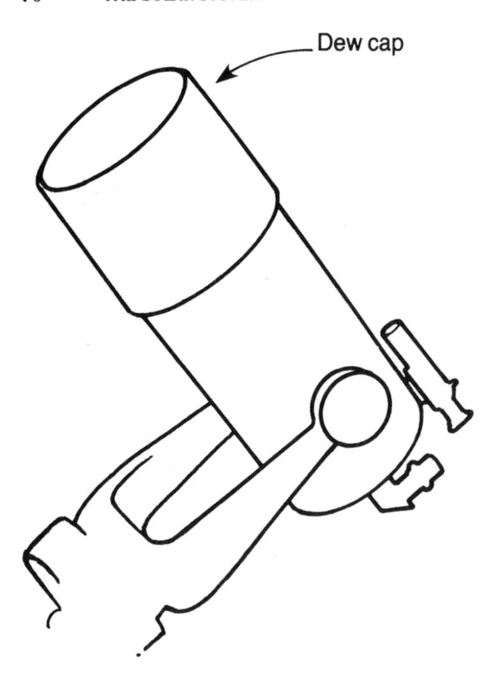

Fig. 6.2. A dew cap on the front of a Schmidt–Cassegrain telescope. The dew cap need be nothing more than a hollow cylinder lined with, or even made of, blotting paper.

shifts, your eye follows it automatically). In any case, a visual observer can make use of fleeting moments of steadiness in a way that a camera can't, which is why you can never photograph as much detail with any telescope as a trained observer can see with the same instrument.

The battle against unsteady air is in some respects a hopeless one; you just have to wait for exceptional conditions if you want exceptional photographs. Don't waste film trying to photograph planets through air that is worse than average; but when the air is steady, take several exposures of each object – from four to a dozen – in the hope of catching a particularly steady moment. You might think that extra-short exposures would be best for capturing fleeting moments of steadiness, but the opposite is the case in practice; many observers find that the integrating effect of an exposure of a couple of seconds can smooth out moment-by-moment variations in the air.

The air is usually steadiest in the late night and early morning hours (before dawn) – times when the temperature is not changing very rapidly. It is worst during daylight (the sun heats the ground and causes thermal currents) and shortly after sunset. Good steadiness is often associated with a high pressure area (anticyclone) that has been in the same place for several days, and with a smaller-than-usual difference between the day's high and low temperatures. A slight haze (one that dims the stars and stops them from twinkling but does not blur them the way high clouds do) is often, oddly enough, a good sign, whereas a brilliantly clear sky – exactly what you'd need if you were observing deep-sky objects – often signals very turbulent air; the more the stars twinkle, the worse it is. It goes without saying that the view is best when the object you're observing is high in the sky, since the amount of air you have to look through is much smaller than for objects near the horizon.

In recent years observers have begun to recognize the importance of the air inside and immediately surrounding the telescope – air whose steadiness is at least partly under your control. To begin with, the telescope should be at the same temperature as the surrounding air; allow 30 minutes for 'cool down' after bringing it out from the warm indoors. Turbulence within the tube usually takes the form of currents that spiral along the inside walls; thus it helps to make the tube oversized and/or square. There is some controversy as to whether the tube should be made of a material that conducts heat; it is generally agreed that heat-conducting (for example, metal) tubes cool down faster, but wooden or phenolic tubes perform better while the cool-down is still going on.

If at all possible, place the telescope on grass or sand rather than pavement, and avoid looking through the air immediately above a roof. Paved areas and roofs heat up during the day and produce thermal currents at night.

Finally, there is dew, an ever-present problem for astronomers in humid climates. One way to keep dew from forming on the front lens of a refractor or catadioptric is to install a *dew cap*, as shown in Fig. 6.2; this reduces the rate at which moisture-laden air can reach the surface of the lens. Dew caps

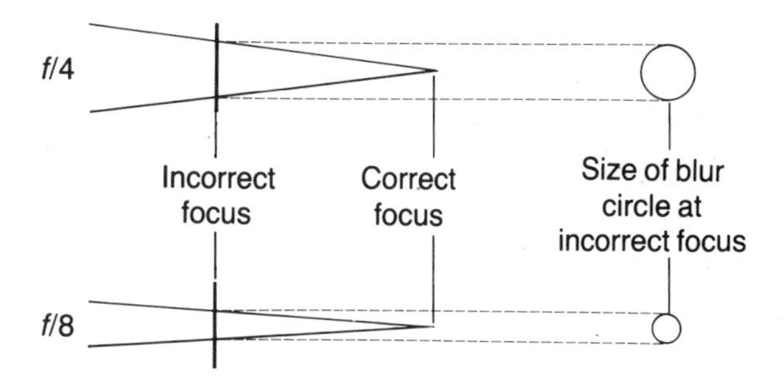

Fig. 6.3. *Effect of f-ratio on blur-circle size.*

up to about 15 cm (6 inches) in diameter can be made out of PVC pipe fittings; for larger ones, a square or hexagonal wooden design is often convenient. To absorb moisture, line the dew cap with a piece of black blotting paper, which you can replace periodically as it becomes damp. In fact, the dew cap can perfectly well consist of the blotting paper by itself, rolled into a cylinder around the end of the telescope tube and secured with an elastic cord. In an emergency, you can make a quite serviceable dew cap out of a brown paper bag.

Some observers use electrically heated dew caps – they place a ring of nichrome wire or carbon resistors around the objective to keep it slightly warmer than the air, thereby avoiding dew. If you do this, use only the minimum amount of heating necessary to prevent dew formation – typically about 2 watts – and use a low voltage for safety's sake.

The design parameters for heating a dew cap electrically are the following. In general, you start out knowing the wattage that you want (for example, 2 watts) and the voltage from which you want to power it (say, 12 volts). The total resistance needed is given by the formula:

$$\text{resistance (in ohms)} = (\text{voltage})^2 / \text{wattage}$$

or, in this case,

$$\text{resistance} = (12)^2 / 2 = 144 / 2 = 72 \text{ ohms}$$

Since you want the heat distributed uniformly around the edge of the lens, you decide to use a string of six 12-ohm resistors in series. (The resistance of resistors in series is equal to the sum of the individual resistances.) The power output is divided equally among the six resistors; each puts out about ⅓ watt, and readily available ½-watt resistors are satisfactory. (The wattage of a resistor represents the maximum it can handle; the actual amount of heat dissipated depends on the voltage and the resistance.)

Finally, in choosing a power source, you need to know how many amperes a 72-ohm resistance will consume at 12 volts. The relevant formula, known as Ohm's Law, is:

$$\text{current (in amperes)} = \text{voltage} / \text{resistance}$$

or, in this case,

$$12 / 72 = 0.167 \text{ ampere} = 167 \text{ milliamperes}$$

This is a relatively small amount of current and could be supplied by an automobile's electrical system, a lantern battery, or a small transformer. It doesn't matter whether AC or DC power is used, but if you use AC, you can cut the total power in half by putting a silicon rectifier in series with the circuit – a convenient way to get switch-selectable 'high' and 'low' settings without having to change the voltage.

A simple way to get rid of dew that has already formed is to blow warm air onto the affected lens with a hair dryer. Try not to overheat the lens, and allow a few minutes afterward for cool-down, if necessary. For portable operation, you can use a 12-volt hot-air blower designed for melting ice on automobile windshields.

The subtle art of focusing

One of the simplest ways to get sharper astronomical photographs is to improve the accuracy of your focusing. Indeed, focusing is a more important – and more difficult – endeavor than many hobbyists realize.

One reason accurate focusing is difficult is that, because of air turbulence and other resolution-limiting factors, you are almost always trying to focus an image that is inherently blurry and which, no matter what you do, will not be crystal-clear. Hence the temptation not to strive for precision.

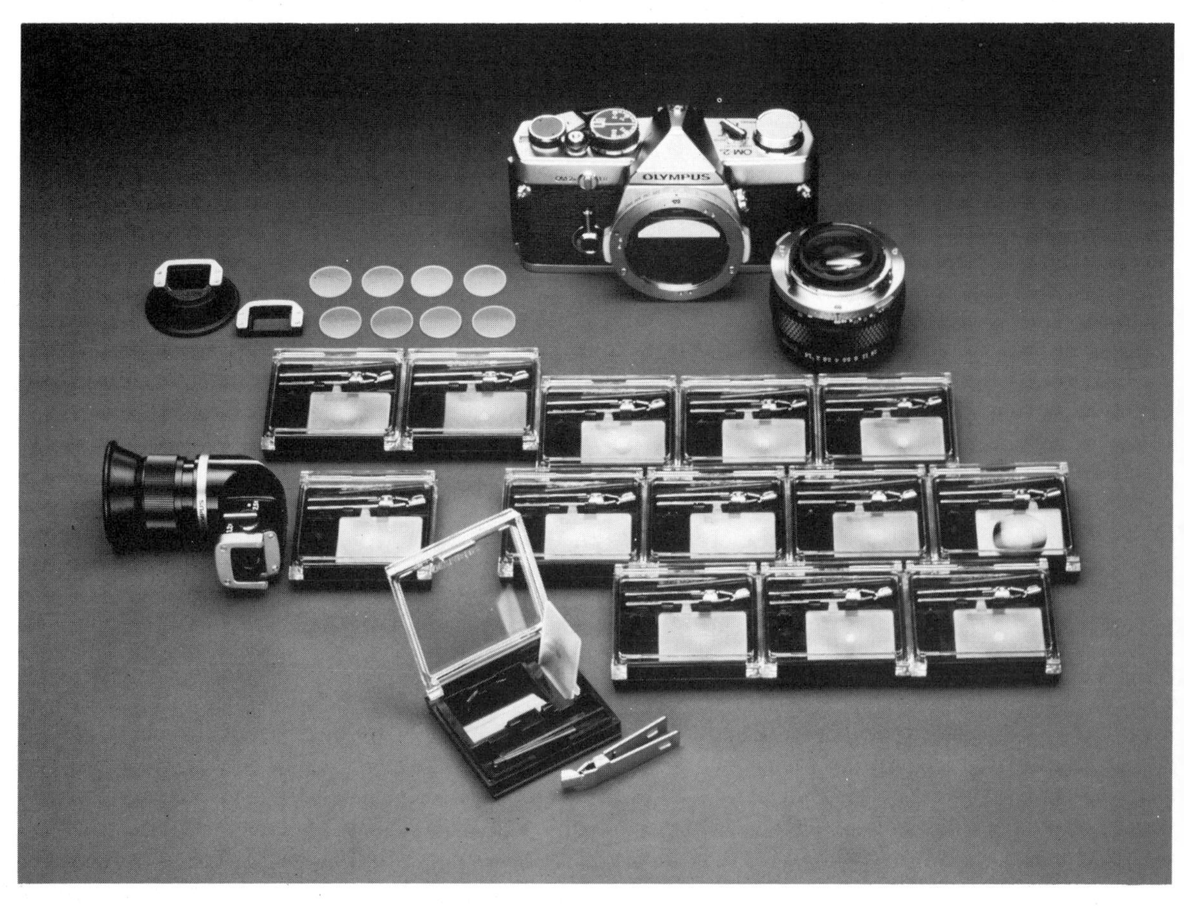

Fig. 6.4. *Focusing aids for the Olympus OM-1 and OM-2. The small plastic boxes contain focusing screens and the tools for inserting them. At their left is a right-angle focusing magnifier. To the left of the camera body are correcting lenses (for eyeglass wearers who prefer to remove their glasses when focusing) and a glare-reducing eyecup. (Olympus Optical Company)*

Another reason is that when you focus either an ordinary camera lens or a visual telescope, you have an advantage that is lacking in an astrophotographic setup. In the case of a visual telescope, the focusing mechanism of your eye will correct automatically for slight errors. And in the case of an ordinary SLR camera, you focus at a wide aperture (say *f*/1.8) and then take the picture at a narrower aperture, so that the effects of any error are greatly reduced (Fig. 6.3), and your picture comes out sharper than what you saw in the viewfinder. With the telescope, you focus at the *f*-ratio at which the picture will be taken; the end result will be no better than what you see at the time. (Incidentally, contrary to what Fig. 6.3 might suggest, an *f*/100 telescope system is no more tolerant of focusing errors than an *f*/4 camera lens. The only way of finding the correct focal plane is by observing the sharpness of the image – which is why all that matters is how small you can get the blur circle, not how far you have to move the focusing tube.)

To make matters worse, the split-image or microprism screen that assists you when you focus your camera doesn't work at *f*-ratios slower than about *f*/4. (Olympus does make one that works down to *f*/8 – screen number 1–2 for its OM-1 and OM-2 cameras – but this is hardly adequate for planetary photography at *f*/120 or the like.) A plain matte ground glass is a bit better; indeed, if your camera does not have interchangeable focusing screens, the fine matte area immediately surrounding the microprism or split-image spot is what you'll have to make do with. Though ideal for bright stars, ground glass is far from satisfactory for low-contrast

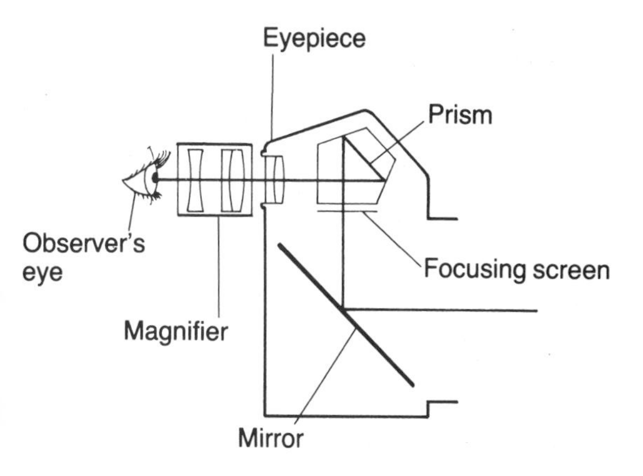

Fig. 6.5. *Use of a focusing magnifier.*

planetary detail; in fact, if you are using a plain matte screen, you may prefer to focus on a nearby star and then aim your telescope at the planet without changing the focus. (Observers of Jupiter are fortunate to have the Galilean satellites for this purpose.) If this is not feasible, focus by concentrating on fine, low-contrast detail that disappears if the focus is anything less than perfect.

The best focusing screen for solar, lunar, and planetary work is a clear piece of glass with crosshair lines scribed on it at the focal plane. Some camera makers manufacture these (an example is the Olympus 1-11, which is advertised as being for photomicrography), or you can sometimes make your own. The technique for using such a screen is simple: keep your eyes focused on the crosshairs and adjust the telescope so that the image is in perfect focus at the same time. As a check, try moving your head slightly; if the focus is perfect, the image will not shift relative to the crosshairs.

The clear-screen-with-crosshairs technique is ideal above about *f*/30; it gives you a brighter, crisper image than any other method, and you need no magnification beyond that provided by the viewfinder eyepiece of the camera. In the *f*/8 to *f*/30 range, however, your eye may not be able to distinguish sufficiently fine differences in focus between the crosshairs and the image. The solution is to add a *focusing magnifier* (Fig. 6.5) – a tiny Galilean telescope that magnifies what you see in the viewfinder. You can make your own out of a pair of lenses that need not be achromatic; the focal length of the concave lens should be about one quarter to one half that of the convex lens, and the spacing should be determined by experiment and left somewhat adjustable. Alternatively, you can buy a focusing magnifier ready-made; for an inexpensive one, inquire at your local camera store or write to Spiratone, Inc. (see Appendix A). Fancier models from camera manufacturers often incorporate such conveniences as right-angle viewing but can cost as much as the camera body itself.

Whichever method you adopt, remember that your focusing will improve with practice. You may want to start each session by focusing your setup several times for practice before making any exposures. When you start taking pictures, remember to focus each exposure individually, and focus all of them with equal care. (I once noticed that the first exposure from each of my sessions was invariably the sharpest; it turned out that the first exposure was the only one I was really doing a good job of focusing.) Many people find that it is better to move through the focusing range quickly rather than slowly; slight changes in sharpness are easier to see if they happen quickly.

There is one last focusing technique worth knowing about because you can use it in situations where nothing else works– *knife-edge focusing*. To use this technique, you do not need a viewfinder or even a piece of ground glass – only access to the film plane before the film is loaded. The results are more precise than those of any other method. The catch is that you can only focus on a star, not a planet or other extended object.

What you do is this: with the telescope aimed at a bright star, open the back of the camera and look in. You should see a large blur circle (with a hole in the middle if the telescope is a reflector or catadioptric). Adjust the aiming to center the blur circle in the field of view. Then take a knife or similar object that you can use to interrupt the light beam precisely at

Fig. 6.6. *An aluminized Mylar sun filter in a home-made wooden mount, on the author's 12.5-cm (5-inch) f/10 Schmidt–Cassegrain telescope.*

the plane of focus, and pass it across the field, through the blur circle. (Use the sharp edge of the knife so that the obstruction does not have any appreciable thickness at the point that matters.) If the blurred star image disappears all at once, you are at perfect focus. If the effect is that of a shadow moving across the image in the same direction as you are moving the knife, rack the camera outward; if the shadow moves in the opposite direction, rack the camera inward.

Incidentally, you can also use the knife-edge test as a way of checking optical quality (the *Foucault test*). If you can't get the whole blurred disk to fade out uniformly, something is wrong. The most common fault in home-made Newtonians is for an area near the edge of the mirror (and hence near the edge of the blur circle) to behave differently from the rest. In a more compelx system, center-versus-edge discrepancies often indicate spherical aberration, and discrepancies between one side and the other (remaining constant as you try bringing the knife in from different directions) indicate misalignment. This test also shows, in minute detail, the effects of air turbulence.

The sun

The basics of solar photography have already been covered in Chapter 3, in connection with eclipses – but there is a lot to see even on the uneclipsed sun. There are almost always a few sunspots (their number varies cyclically, with a maximum in 1990, a minimum around 1997, and a maximum again in 2001); on a good day you can also see *faculae* (bright hydrogen patches visible near the edge of the disk), and the surface of the photosphere shows a low-contrast granular pattern. A telescope of 10 or 12.5 cm (4 or 5 inches) aperture, with an aluminized Mylar filter in front, is the ideal instrument (Fig. 6.6); because the daytime air is inherently unsteady, larger apertures cannot be used except on mountaintops – which means, at least, that the amateur's modest telescope is likely to be as good as any.

The main challenge is to keep the sun from heating the telescope and objects around it, causing severe air turbulence. Be sure to keep the telescope aimed straight at the sun as much of the time as possible, to keep sunlight from falling on the outside of the tube and causing irregular heating. A white tube is, of course, less vulnerable than a dark-colored one.

It is hard to predict at what time of day the air will be steadiest. Some observers prefer to work in the early morning and, surprisingly, just before sunset; although the light has to pass through a greater thickness of air then than at midday, there is less turbulence due to heating than in the middle of the afternoon. However, this generalization does not hold up for all observing sites; a lot depends on what is within a few hundred meters of the telescope in the relevant direction. Around Los Angeles, heavy smog seems to help steady the air (surely its only known benefit to mankind). Any amount of cirrus cloud, however, blurs the sun considerably.

The film should be very fine-grained so that if you succeed in recording the granulation of the photosphere, you will not confuse it with film grain; contrast should be somewhat higher than in pictorial photography, though not excessively high. Kodak Technical Pan and T-Max 100 are the films of choice. There is little point in using color film, since the filter usually alters the color dramatically.

Monochromatic filters that make it possible to photograph flares and prominences are now within

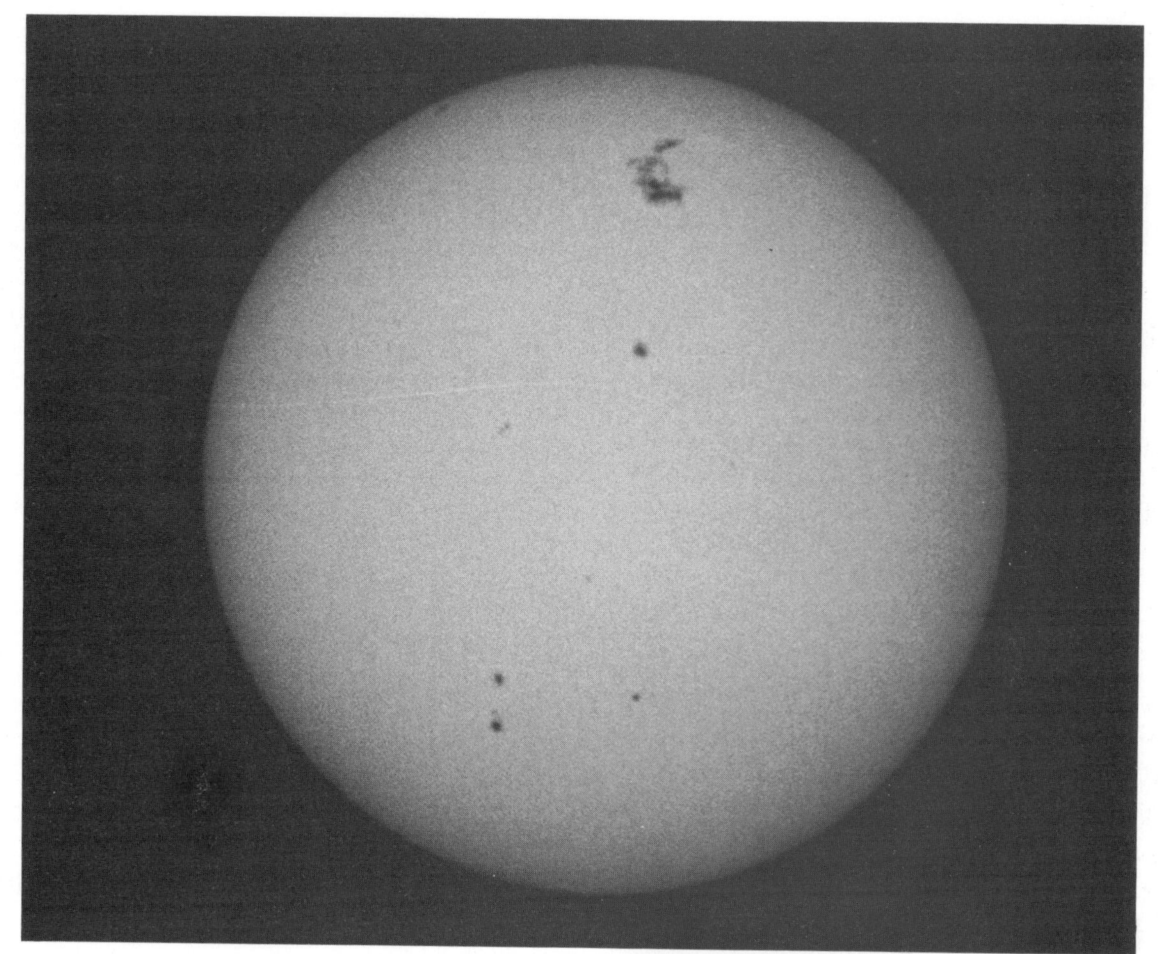

Fig. 6.7. The sun, showing a large sunspot group that had caused magnetic storms on earth a few days before. Taken with the apparatus shown in Fig. 6.6, 1/60 second at prime focus on Agfa Superpan (ISO 200). (By the author)

Fig. 6.8. A moderately high-resolution view of the solar eclipse of 30 May 1984. Taken by the author with a 12.5-cm (5-inch) Schmidt–Cassegrain telescope, 1/60 second on Technical Pan 2415 developed 6 minutes in HC-110(D) at 20 °C (68 °F). Note the faculae (lower left), the large sunspot group, and the jaggedness of the moon's limb (near the sunspots).

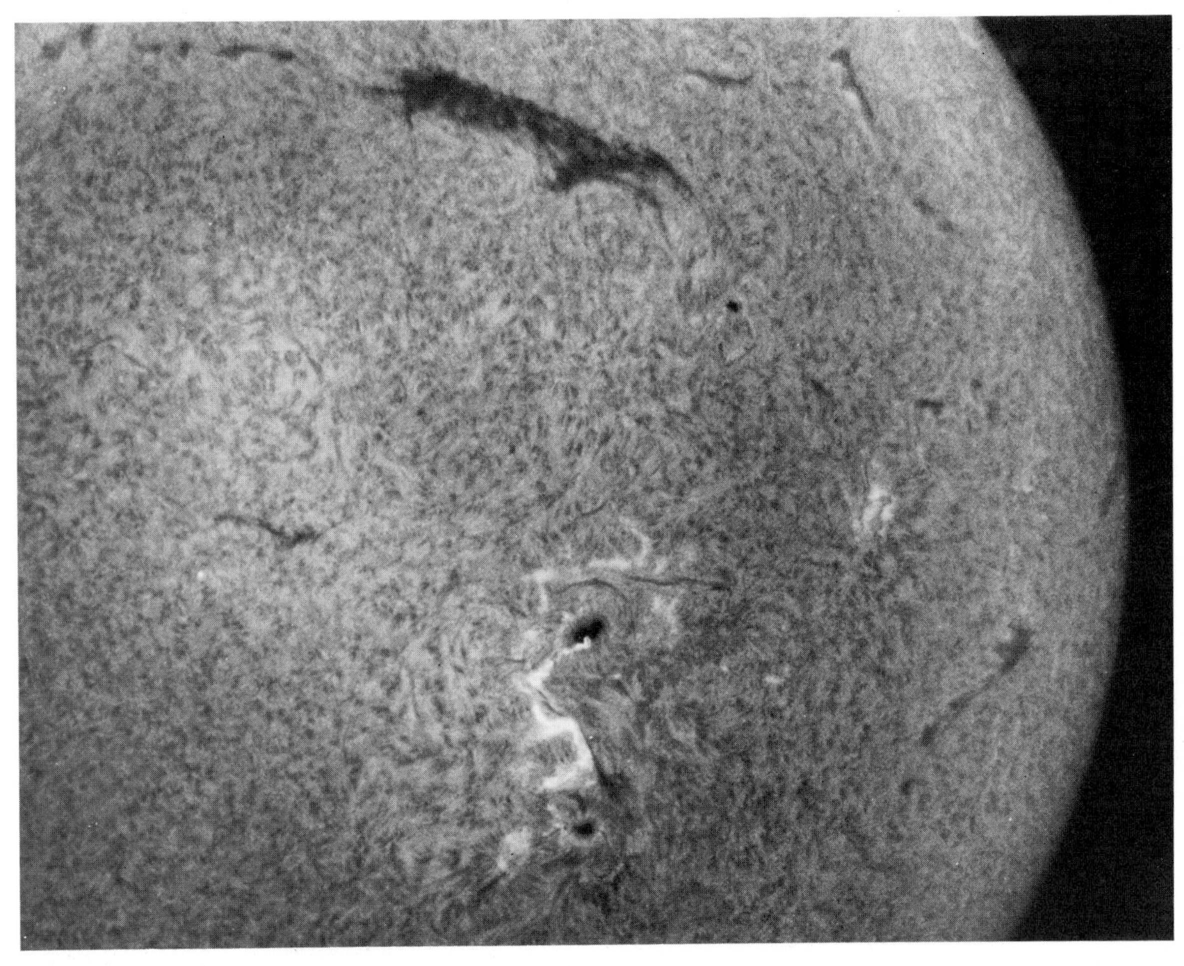

Fig. 6.9. *The sun through a narrow-bank hydrogen-alpha filter. (Edwin Hirsch)*

the reach of at least some amateurs, and amateur photographers of this type can rival the work of observatories. The idea is to use a filter that transmits only light of the wavelength produced by a particular electron transition in a particular chemical element – usually the 'hydrogen alpha' wavelength of 656.281 nanometers – so that you can see the activity of that element while greatly attenuating everything else. The cheaper wide-band filters allow photography of prominences only at the edge of the solar disk, but the more efficient narrow-band filters reveal hydrogen regions all over the sun. The telescope works at *f*/30 to *f*/60 and need not be achromatic; a few observers have had success mating a $1000 filter with a $5 plano-convex lens, though most use a regular telescope stopped down to about 5 cm (2 inches) aperture. The film of choice is Kodak Technical Pan; the 656-nm wavelength is too far off the red end of the spectrum for most other films. Fig. 6.9 shows the results that can be obtained. For more information, see Edwin Hirsch, 'Photographing the sun in hydrogen alpha', *Astronomy*, January 1978, pp. 36–9, or write to Edwin Hirsch (Telescopes and Filters), 168 Lakeview Drive, Tomkins Cove, NY 10896.

The moon

Lunar photography has already been covered in some detail in Chapter 2, and the only technical point that remains to be covered is tracking. Most clock drives are built to make one revolution in exactly 24 hours. This allows the use of standard electric clock parts, and it gives exactly the tracking

Fig. 6.10. *The gibbous moon as photographed by Dennis Milon at the prime focus of a 40-cm (16-inch) f/6 reflector, on Technical Pan 2415 film. Although not quite as sharp, photographs taken with telescopes in the 12.5- to 20-cm range look similar.*

rate you need for the sun; it is off by only 1° per day for the stars, and usually by somewhat less for the planets. The moon, however, orbits the earth so rapidly that its apparent motion in right ascension differs from the solar rate by some 10° to 15° per day, or 0.5 arc-second per second of time, and special procedures are necessary for tracking it.

If you can keep the total tracking error under 0.5″, you can ignore it – it will be negligible relative to the blurring introduced by the earth's atmosphere, regardless of the size of the telescope. This means that, with a solar rate drive, your lunar exposures should be less than one second long. This is not a very restrictive constraint; the moon is bright enough that exposures of less than one second are almost always feasible.

For longer exposures, you can track at the moon's own rate by making your drive run at only 97% to 98% of its usual speed. This is done by supplying it with electric power, not from the electric mains, but from a *drive corrector* – a device that allows you to vary the frequency of the alternating current, and hence the motor speed (see Chapter 7 and Appendix D). Aim your telescope at the moon, view through a high-power eyepiece or one with crosshairs, and adjust the drive corrector until there is no noticeable error over a period of several minutes. There is no point in trying to isolate a single, precise 'lunar rate', since the moon's daily motion in right ascension varies widely depending on its position in its orbit. There is also motion in declination – as much as 5° per day – which the drive can't help you with.

Don't be surprised if you find that even the most finely adjusted drive corrector doesn't give you perfect lunar tracking. Once you get down to the arc-second level, even the best-built clock drives are full of inaccuracies and periodic errors (again, see Chapter 7). The moral is that, for highest resolution, you shouldn't trust your drive for more than a few seconds.

The planets

It is in planetary photography that you really need magnification. Fig. 6.12 shows the comparative apparent sizes of the planets as seen from earth. The size of the image on the film (w) depends on the angular diameter of the planet in arc-seconds (d) and the effective focal length of the telescope (F):

$$w = (F \times d)/206265$$

Thus, for example, a 200-mm (8-inch) diameter telescope operating at $f/64$ has an effective focal length of 12 800 millimeters; photographing Jupiter (apparent diameter, about 40″) with such a telescope, you get an image about 2.5 mm across on the film. This can comfortably be enlarged ×10 to make the image 25 mm (1 inch) in diameter on the print, big enough to show all the detail a 20-cm telescope will normally capture.

If you had been photographing Mars at 20″ apparent diameter, you would have needed a ×20 enlargement to give the same image size – more enlargement than an ordinary enlarger will give. One solution to the problem is to turn the enlarger backward on its support and project onto the floor. Another is to use a short-focal-length enlarging lens designed for 110 film; you do not have to cover the entire 35-mm frame, of course, but make sure your enlarger will let you get the extra-short lens close enough to the film to focus it.

And when you make ×20 enlargements, film grain becomes a significant problem. The ideal solution is to use a film such as Technical Pan 2415, which can withstand ×20 enlargement without difficulty. Alternatively, if you want to make a big enlargement of a rather fuzzy image on a film like Tri-X, you can simply put the enlarger slightly out of focus; the slight loss of resolution in the image will be more than made up for by the elimination of distracting grain. (This is essentially how a chromogenic film such as Ilford XP-1 gets rid of apparent grain: the microscopic silver grain clumps are replaced by fuzzy-edged blobs of dye, so that although the resolution is more or less unaffected, the amount of visible grain is reduced substantially.)

A much more subtle solution is to combine several negatives into one print, either by sandwiching them or, more commonly, by multiple exposure. The idea here is that details common to all the negatives – representing actual features on the planet – will reinforce, while variations from one image to another, comprising atmospheric distortion and film grain, will cancel out. Fig. 6.14 shows the result.

Applying this technique in practice can be rather tricky. One way of doing this is to have a piece of ordinary paper which you can substitute for the photographic paper in the enlarging easel, and on which you can make pencil marks. After making the first exposure, put the photographic paper away in a light-tight place, put the plain paper in the easel, and pencil in the outline of the planet. Now take the first negative out of the enlarger, put in the second one, position it to match the penciled outline, and refocus. Then put the photographic paper back in the easel and make the second exposure. These steps can be repeated for as many exposures as necessary. Naturally, the exposure time should be reduced in proportion to the number of negatives being combined, so that if, for example, you are combining five negatives, each of them will receive only one-fifth the exposure time that would be correct for a single negative by itself.

Planetary photography often involves the use of filters, and Table 6.1 lists all the filters you are ever likely to come across, including a good many that are not particularly useful. The purpose of a filter is to block light of certain wavelengths while transmitting others; in general, the effect on the picture is to lighten features whose colors fall within the transmitted range while darkening those whose colors are blocked. For example, a yellow filter

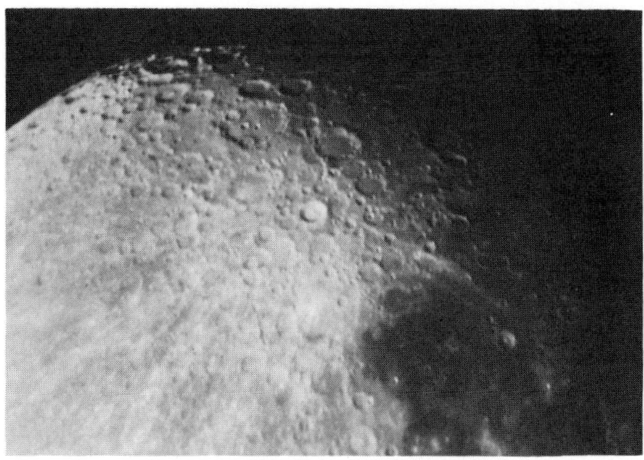

Fig. 6.11. *The crater Tycho and its ray system, photographed with a 12.5-cm (5-inch) f/10 Schmidt–Cassegrain telescope and a ×3 teleconverter. 1/4 second on Kodak SO-115 (now known as Technical Pan 2415) developed 6 minutes in HC-110(E) at 20 °C (68 °F). (By the author)*

Fig. 6.12. *Relative sizes of the planets as seen from earth.*

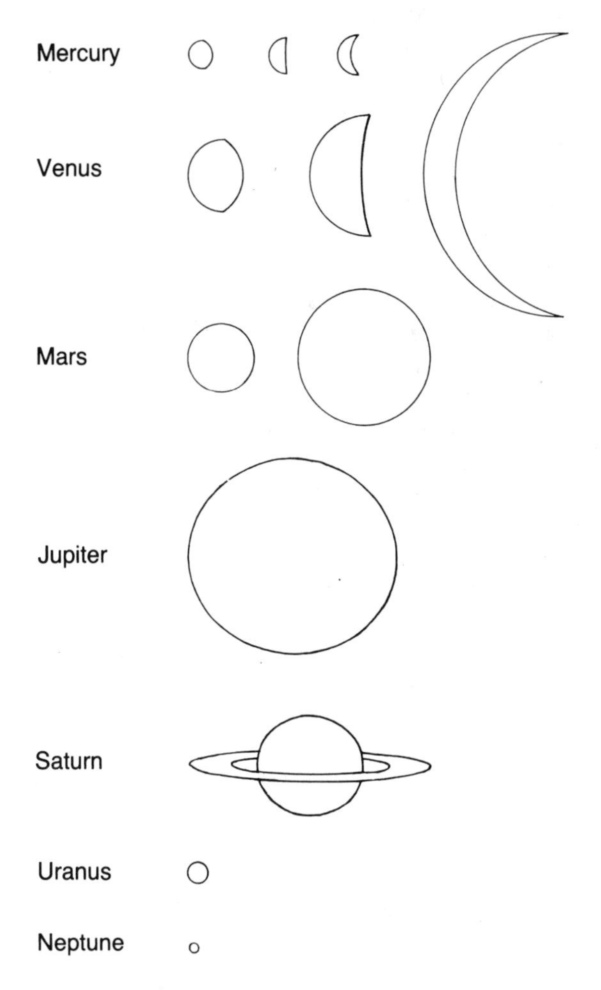

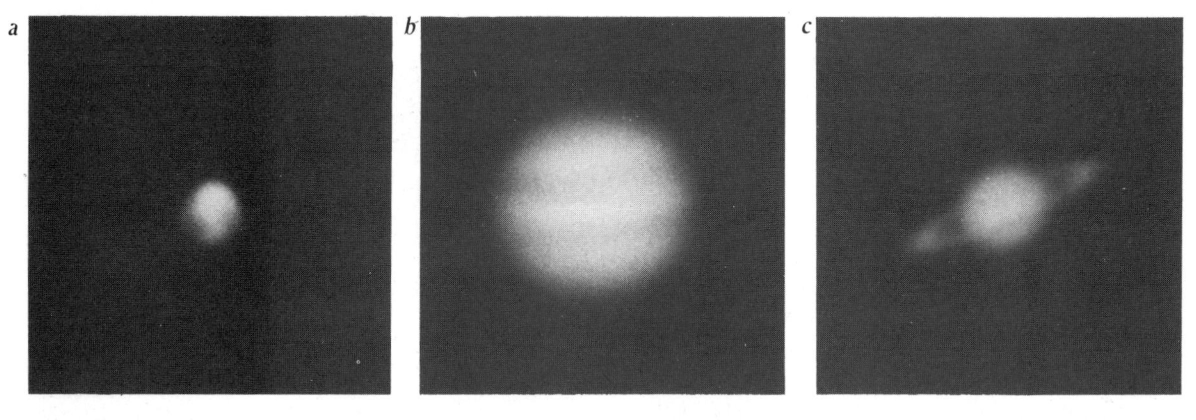

Fig. 6.13. *Three planets as photographed by the author with a 12.5-cm (5-inch) f/10 Schmidt–Cassegrain with an 18-mm eyepiece coupled afocally to a 100-mm telephoto lens for an effective focal length of 7000 mm at f/56. All on Kodak SO-115 film (now known as Technical Pan 2415) developed 6 minutes at 20 °C (68 °F) in HC-110(D). (a) Mars, 1 second, no filter (polar cap at top); (b) Jupiter, #15 yellow filter, 3 seconds; (c) Saturn, no filter, 7 seconds.*

lightens yellow and red, but darkens blue and violet.

Filters that screw into eyepieces and Barlow lenses are readily available, and some catadioptric telescopes are made to take readily available filters of standard sizes. Also, the old Series 5 drop-in filters, now available very cheaply from camera stores that accumulated large stocks of them thirty years ago, are just the right size to fit inside a 32-mm (1¼-inch) diameter eyepiece tube if you can improvise a retaining ring to hold them in place.

Since the filter affects the total amount of light available, the exposure has to be corrected to allow for it. The best way to do this is to divide the ISO (ASA) speed of the film by the filter factor (from Table 6.1), then use the corrected film speed when consulting the exposure tables:

corrected speed = original speed / filter factor

Remember that filter factors are only approximate, and bracket exposures widely.

Now for the planets themselves. Mercury, which varies from 4.8″ to 13.3″ in apparent diameter, is rather hard to photograph because it is so close to the sun. If you want to observe Mercury when the sun is not in the sky, you have to look very close to the horizon right after sunset or before sunrise – which means you're looking through a thick layer of turbulent air, and the prospects of seeing any surface detail are poor. Since much of the blurring introduced by the air, especially near the horizon, consists of chromatic aberration, anything that restricts the range of wavelengths you use, such as a red, green or blue filter, will help. Alternatively, you may choose to observe in the daytime, using a deep red filter to suppress the blueness of the sky.

Venus is a much easier target. It has the highest surface brightness of any planet and, at times, the largest apparent diameter (64″). Recording the phases of Venus photographically is an excellent project for the beginner (Plate 6.1).

Photographing detail on Venus is more of a challenge – it requires the use of an ultraviolet filter (the Wratten #18A) that transmits no visible light, so that you can't see to focus. The solution is to focus through a different-colored filter made of the same type and thickness of glass, then swap filters before making the exposure. (By the way, the term 'ultraviolet filter' is ambiguous – you want the almost-black #18A not the clear #1A.) Ordinary film is quite sensitive at ultraviolet wavelengths, but not all telescopes are suitable for ultraviolet work; most refractors suffer severe chromatic aberration in the ultraviolet. For the same reason, complicated afocal or projection systems are a bad idea – the lenses in them won't really be achromatic at the wavelengths you're using. In fact, the amount of glass in the system should be held to a minimum, since all glass absorbs ultraviolet to some extent; it is best to work at the prime focus of a reflector.

Mars is, in my opinion, somewhat neglected as an

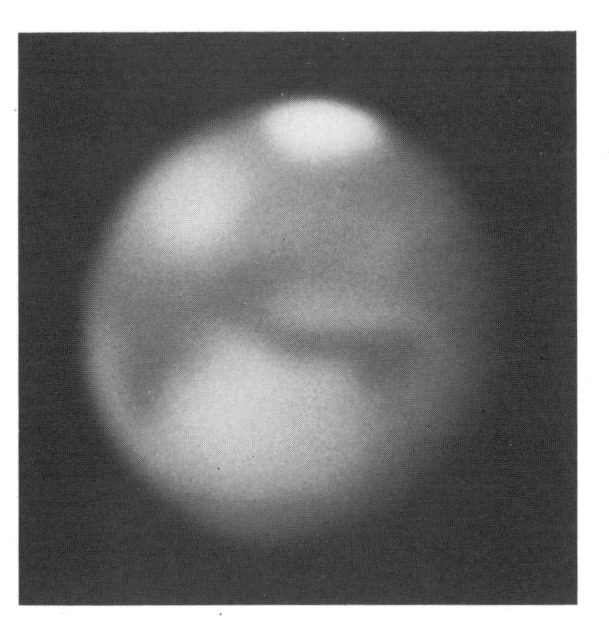

Fig. 6.14. Mars, photographed by Dennis Milon at the Lines Observatory, Mayer, Arizona, with a 40-cm (16-inch) f/8 Newtonian using positive projection to give f/235. Six Tri-X negatives were combined by copying onto Kodak Contrast Process Ortho Film; the resulting positive was converted into a negative with Kodak Professional Copy Film.

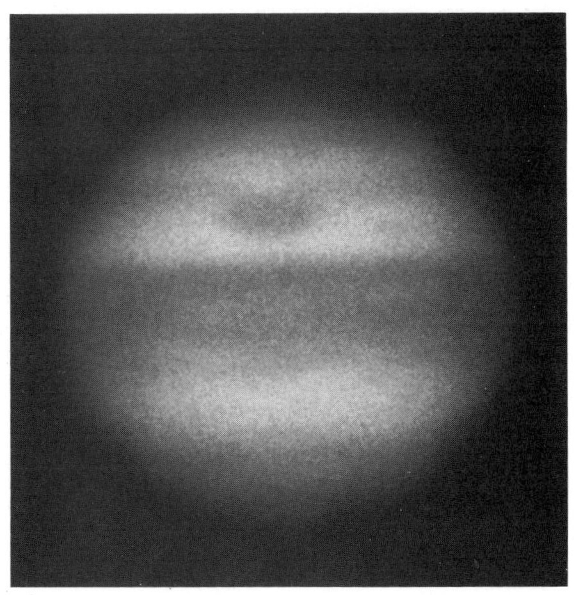

Fig. 6.15. Dennis Milon's photograph of Jupiter, taken with a 20-cm (8-inch) reflector, is a good example of high-resolution astrophotography. The picture resembles the visual appearance of the planet in a telescope perhaps half the size.

astrophotographic target; it is in fact the easiest planet on which to photograph detail because of the relatively high contrast of the surface features. A red filter helps bring out surface detail; a green filter is useful for highlighting the polar caps by darkening everything else. Good color photographs (with no filter, of course) are not hard to obtain.

The apparent diameter of Mars does vary quite a bit with its position in its orbit. The best observing conditions in the near future will occur in the late summer and autumn of 1988, at which time the apparent diameter will peak at 23.8″; from then on, each opposition will be worse than the previous one until that of 1995 (13.9″), after which the situation will improve again.

Jupiter (31″ to 48″) and Saturn (15″ to 21″) are classic targets for the amateur astrophotographer; surface detail, though low in contrast, is abundant, and even a 10-cm (4-inch) telescope will show a couple of bands on Jupiter. A yellow filter is generally helpful, as is high-contrast film (Technical Pan 2415 developed in HC-110(D) or the like). Take advantage of exceptional atmospheric conditions when they occur, and try color film, especially when the Great Red Spot is prominent.

Fig. 6.16. This long exposure shows the trail of the asteroid Eros moving relative to the stars. Thirty minutes on Fujichrome 100 at the prime focus of a 12.5-cm (5-inch) telescope. (George East)

Uranus (4″), Neptune (2.5″), and Pluto (0.1″) have little to offer the amateur observer. You can, however, treat them as stars (of sixth, eighth, and fifteenth magnitude, respectively) and include them in star-field pictures. Asteroids can be handled similarly; they often move fast enough to leave a trail (Fig. 6.16).

Table 6.1. *Commonly encountered photographic filters*
*'Wratten' is a Kodak trademark, but the numbering system has been adopted by many other manufacturers. Note that most filters transmit infrared – a fact that you can ignore unless you are using infrared film. Filters marked * are useful in astrophotography; those marked **, particularly so.*

Wratten number	Old Wratten letter	Filter factor		Visual appearance	Remarks
		Most films	Kodak Technical Pan		
*1A	–	1.0	1.0	Clear ('Skylight')	Blocks ultraviolet (below 390 nm), giving slight reduction of haze or sky fog. Transmits all visible wavelengths
*3	–	1.5	1.2	Light yellow	Blocks ultraviolet, violet, and deep blue (below 465 nm)
6	K1	1.5	1.2	Light yellow	Partly blocks ultraviolet, violet, and deep blue; weaker version of #3
*8	K2	2	1.5	Yellow	Blocks blue, violet and ultraviolet (below 495 nm)
*9	K3	2	1.5	Yellow	Like #8 but blocks some blue-green as well – cutoff point is 510 nm
11	X1	4	4	Yellow-green	Blocks ultraviolet, violet and deep blue (below 480 nm); partly blocks orange and red (above 580 nm)
**12	–	2	1.5	Deep yellow	Blocks ultraviolet, violet, blue, and some green (below 520 nm). Cutoff is very sharp; transmission above 520 nm is about 90%
*13	X2	5	5	Dark yellow-green	Transmits green and yellow (480–580 nm). A more efficient version of #11
**15	G	2.5	2	Deep yellow	Very similar to #12; you won't need both. Cutoff is a bit different (530 rather than 520 nm)
*18A	–	–	–	Black (visibly opaque)	Transmits ultraviolet (320–380 nm); blocks all visible light. Transmits a small amount of infrared
*21	–	4	2	Orange	Blocks ultraviolet, violet, blue, and some green (below 555 nm)
*23A	–	6	2	Light red	Blocks ultraviolet, violet, blue, and green; transmits infrared, red, and orange (above 580 nm)
**25	A	8	3	Red	Transmits only infrared, red, and some orange (above 600 nm)
*29	F	20	8	Deep red	Transmits only infrared and red (above 620 nm)
34A	–	8	~8	Purple	Blocks yellow, green, and some blue (455–650 nm); transmits red and most of blue and violet

Wratten number	Old Wratten letter	Filter factor		Visual appearance	Remarks
		Most films	Kodak Technical Pan		
*44	–	8	~12	Blue-green	Transmits blue and green (450–535 nm) and infrared (above 730 nm)
**47	C5	6	12	Blue	Transmits violet and blue (410–480 nm) and infrared (above 760 nm)
*47B	–	8	16	Deep blue	Transmits violet and blue (410–460 nm) and infrared (above 750 nm). A darker version of #47; you won't need both
49	C4	–	–	Dark blue	Partly transmits blue (430–470 nm); transmits infrared (above 750 nm)
**58	B	8	12	Green	Transmits green (505–560 nm) and infrared (above 725 nm)
61	N	12	20	Dark green	Partly transmits green (550–560 nm); transmits infrared (above 750 nm)
80A, 80B	–	4	~8	Pale blue	Slight absorption of green, yellow and red. Intended for adjusting color rendition of color films; of little use in black-and-white
80C	–	2	~4	Pale blue	As 80A, 80B
81, 81A, 81B, 81C	–	1.2	1.0	Pale yellow	Some absorption of blue. Intended for adjusting color rendition of color films; of little use in black-and-white
82, 82A	–	1.2	1.5	Pale blue	As 80A, 80B
85, 85B	–	1.5	1.2	Pale orange	As 81, 81A
*87A, 87B, 87C, 88A, 89B	–	–	–	Black (visibly opaque)	Transmits infrared only
*92	–	–	–	Very dark red	Transmits only infrared and extreme red (above 650 nm). Compare to #29
**96	(ND)	–	–	Neutral gray	Available in many different densities. Absorbs equal amounts of all visible wavelengths. Transmits more infrared than visible light; transmits little ultraviolet

7
Deep-sky photography

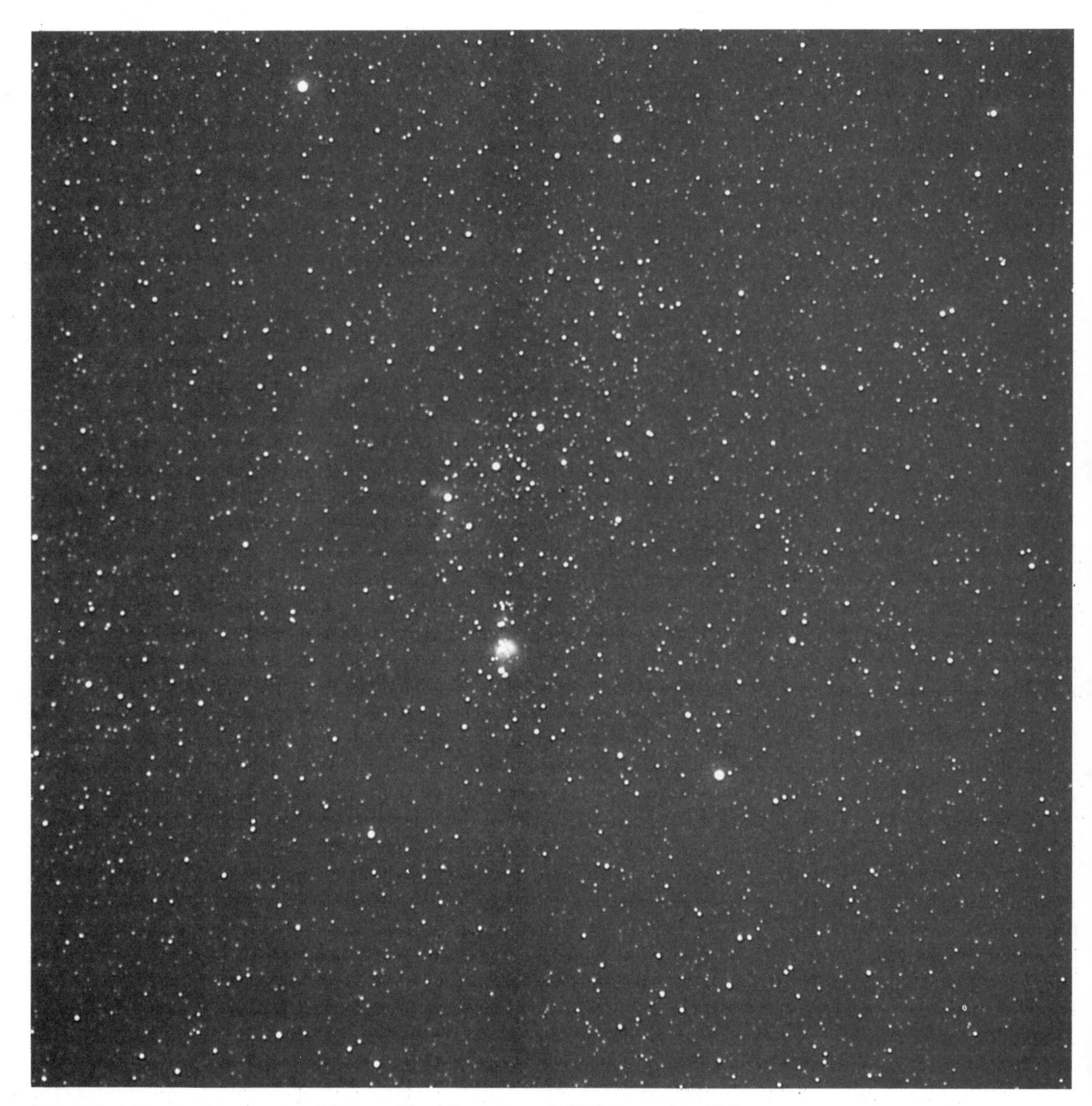

Fig. 7.1. *A 10-minute exposure of Orion with a 50-mm lens at f/2.8 through a #12 yellow filter, on gas-hypersensitized Ektachrome 200. The Orion Nebula and the nebulosity around Zeta Orionis (marked ζ on Fig. 7.2) are easy to see; Barnard's Loop was visible on the original slide, though it may not show up in the reproduction. (By the author)*

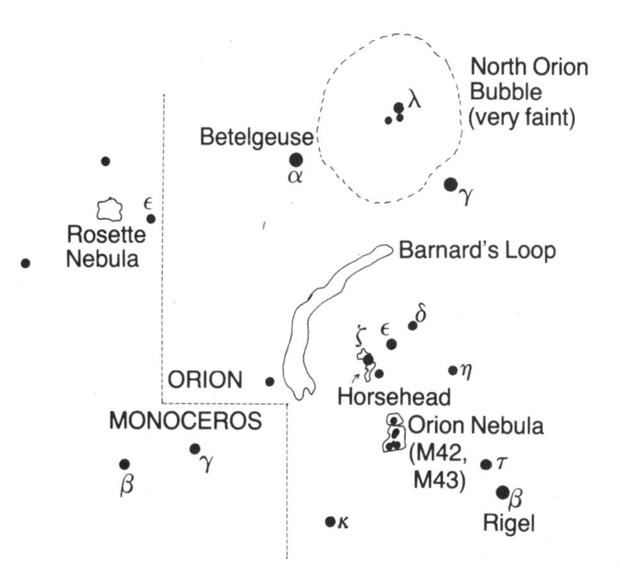

Fig. 7.2. Nebulosities to look for in your photographs of the Orion region. All are fluorescent gas; Barnard's Loop and the North Orion Bubble are probably supernova remnants.

Fig. 7.3. How piggy-backing is done. The camera can be attached to any part of the telescope or even to a counterweight. If guiding corrections are to be made, the telescope and camera should point in roughly the same direction, though not necessarily at the same object.

Beyond the solar system

The 'deep sky' is the sky beyond the solar system – comprising stars, clusters, nebulae, and other galaxies. Many deep-sky objects are extraordinarily beautiful (see, for example, Plates 7.1 and 7.2). Moreover, because of its ability to collect faint light through long time exposures, the camera has the advantage over the human eye in deep-sky work, and a relatively modest camera can photograph objects so faint that no human eye has ever seen them. Professional observatories nowadays do almost all their deep-sky work photographically, a tradition that goes back to E. E. Barnard's discovery of the Loop Nebula, a supernova remnant in Orion, with an inexpensive plate camera in 1894.

Table 7.1 lists some deep-sky objects suitable for photography with modest equipment. Many more are easily located using books such as Burnham's *Celestial Handbook* and Tirion's *Sky Atlas 2000.0* (see Appendix A), or by simply taking a wide-angle photograph of a starry region of the Milky Way.

Piggy-backing

The simplest way to do deep-sky photography is to mount your camera 'piggy back' on a telescope that has an equatorial mount and clock drive (Fig. 7.3). The telescope is used for guiding, but its optical system is not involved in the photography, which is done through the camera lens alone. You can use a wide-angle lens for wide panoramas of sky, a normal lens for single constellations, or a telephoto lens for smaller areas such as M31, the Orion Nebula, or the Pleiades. Fig. 7.4 and Plates 7.3–7.5 show typical results.

The two lens characteristics that matter most are speed and focal length. The light-gathering ability of a lens is determined differently for photographing stars than for photographing extended objects (nebulae, the Milky Way, planets, or terrestrial objects). When you photograph a star, all that matters is the total amount of light that the lens collects, which in turn depends on the diameter of the entrance pupil (roughly, the front opening) of

Name	Type of object	Constellation	RA	Decl.	Size	Brightness[1]
Pleiades	Large open cluster with faint nebula	Taurus	03.44	+30°	2°	Bright
Coma star cluster	Loose cluster	Coma Berenices	12.20	+28°	10°	Bright
Double Cluster	2 open clusters	Perseus	02.17	+57°	1°	Bright
Omega Centauri	Globular cluster	Centaurus	13.24	−47°	0.5°	Bright
Rosette Nebula	Cluster with nebula	Monoceros	06.30	+5°	1.2°	Faint
M42	Bright nebula	Orion	05.24	−5°	2°	Bright
Barnard's Loop	Nebula	Orion	05.50	+2°	12° long	Faint
North America Nebula	Nebula	Cygnus	21.00	+45°	3°	Medium
Magellanic Clouds	Galaxies	Tucana Dorado, Mensa	00.50 05.25	−73° −68°	5° 9°	Bright
M31	Galaxy	Andromeda	00.40	+41°	3°	Bright
M33	Galaxy	Triangulum	01.31	+30°	1°	Medium
Galactic center	Dense Milky Way area with many clusters and nebulae	Sagittarius	17.43	−29°	20°	Bright
Great Rift	Dark lane in Milky Way	Cygnus, Aquila	19.30	+20°	40° long	Bright

Table 7.1. *Some interesting objects for piggyback astrophotography*
This is just a selection – there are many others that are no more difficult.

the lens; all the light from a single star will go into a single point, and the focal length will have no effect on it. If you wanted to magnify a star into anything more than a point, you would need a lens system larger than the largest telescopes on earth.

When you photograph an extended object, on the other hand, you have to deal not only with the amount of light collected, but also with the area over which it is spread out, which in turn depends on the focal length. If you have two lenses of equal diameter but different focal lengths, the longer lens, in forming a larger image, will spread the light thinner and make the image fainter. The image brightness depends on the *f*-ratio, which is the focal length divided by the diameter.

The upshot of all this is that two lenses of the same diameter but different focal lengths, aimed at the same place in the sky for exposures of equal

ested	Month in which best seen[2]	Visible from		
		UK	USA	Australia
hoto	December	Yes	Yes	Yes
nal or ephoto	April	Yes	Yes	Yes
telephoto	November	Yes	Yes	No
hoto	May	No	No	Yes
photo	January	Yes	Yes	Yes
photo	January	Yes	Yes	Yes
normal	January	Yes	Yes	Yes
nal or ephoto	September	Yes	Yes	Yes
	October January	No	No	Yes
photo	October	Yes	Yes	Poorly
photo	November	Yes	Yes	Yes
nal or ide-angle	July	Poorly	Yes	Yes
nal or e-angle	August	Yes	Yes	Yes

[1] Bright = easily visible to naked eye; medium = easily photographed under good conditions; faint = photographable with some difficulty.
[2] At 10 p.m. local mean time.

length on the same kind of film, will photograph equal numbers of stars (within a given area), but the shorter lens, being faster in the *f*-ratio sense, will be more sensitive to nebulae, the Milky Way, and sky fog. (It will also yield a richer-looking picture, since the same number of stars will be compressed into a smaller area on the film.)

On the other hand, two lenses operating at the same *f*-ratio but with different focal lengths, and hence different diameters, will record different numbers of stars while responding to extended objects identically. To be specific, assuming sky fog is negligible, the difference in the stellar magnitude limits for the two lenses will be:

$$\text{magnitude difference} = 2.5 \log_{10} (D_1 / D_2)^2$$

where D_1 and D_2 are the respective diameters. The computed difference should be added or subtracted to the magnitude limit for one lens to obtain that of the other.

The effective diameter of a lens is not necessarily the same as the size of the front element and should always be computed from the *f*-ratio, as follows:

$$\text{diameter} = \text{focal length} / f\text{-ratio}$$

Table 7.2 gives, for comparison, the diameters of many commonly used lenses, and Table 7.3 gives some estimated comparative magnitude limits.

Fig. 7.5 shows the relative merits of various commonly encountered lenses for stars and for extended objects. Remember that most of the really interesting deep-sky objects are extended: nebulae, galaxies, the Milky Way, globular clusters, and so forth. Hence *f*-ratio is usually more important than diameter. The ability to photograph the stars themselves comes into play if you are interested in open clusters, novae, variable stars, or asteroids.

The type of objects you will want to photograph will be determined, of course, by the field of view, which depends on the focal length and film size. The formulae for computing field of view were given in Chapter 5; Table 7.4 summarizes the results for commonly used focal lengths and 35-mm film. (For focal lengths over 500 mm, see Table 5.2 in Chapter 5.) You may want to make an overlay for your star atlas that shows the coverage of each of your lenses, as an aid in planning exposures and comparing the resulting pictures to the charts.

a

Fig. 7.4. *Two photographs with a 200-mm lens at f/3.5, both on Ektachrome 400. (a) The Orion Nebula, 20 minutes; (b) the Pleiades, 5 minutes. (By the author)*

Fig. 7.5. *Relative merits of various lenses for photographing stars and extended objects. '500/8' means '500-mm f/8' and so forth.*

Table 7.2. *Diameter (in mm) of various commonly used lenses*

Focal length (mm)	f/	Diameter (mm)
18	3.5	5
21	2	10.5
21	2.8	7.5
21	3.5	6
24	2	12
24	2.8	8.5
28	2	14
28	2.8	10
28	3.5	8
35	2	17.5
35	2.8	12.5
35	3.5	10
50	1.4	36
50	1.8	28
50	2	25
50	3.5	14
70	4	17.5
85	2	43
85	2.8	30
85	3.5	24
100	2.8	35
100	3.5	29
135	2.8	48
135	3.5	39
180	2.8	64
180	3.5	51
200	3.5	57
200	4	50
200	5	40
300	4	75
300	5.6	54
400	6.3	64
500	8	63

Table 7.3. *Expected magnitude limit in a 5-minute exposure on ISO 200 film (assuming a 50-mm f/2 lens reaches magnitude 10.0)*

Lens	Approximate diameter	Magnitude limit
400-mm f/5.6 300-mm f/4 200-mm f/2.8	70 mm	12.2
300-mm f/5.6 200-mm f/4 135-mm f/2.8 100-mm f/2	50 mm	11.5
200-mm f/5.6 135-mm f/4 100-mm f/2.8 75-mm f/2	35 mm	10.7
100-mm f/4 75-mm f/2.8 50-mm f/2	25 mm	10.0
75-mm f/4 50-mm f/2.8 35-mm f/2	18 mm	9.3
50-mm f/4 35-mm f/2.8 28-mm f/2	13 mm	8.6
35-mm f/4 28-mm f/2.8 24-mm f/2.8	9 mm	7.8
28-mm f/4 24-mm f/4 18-mm f/2.8	6.5 mm	7.0

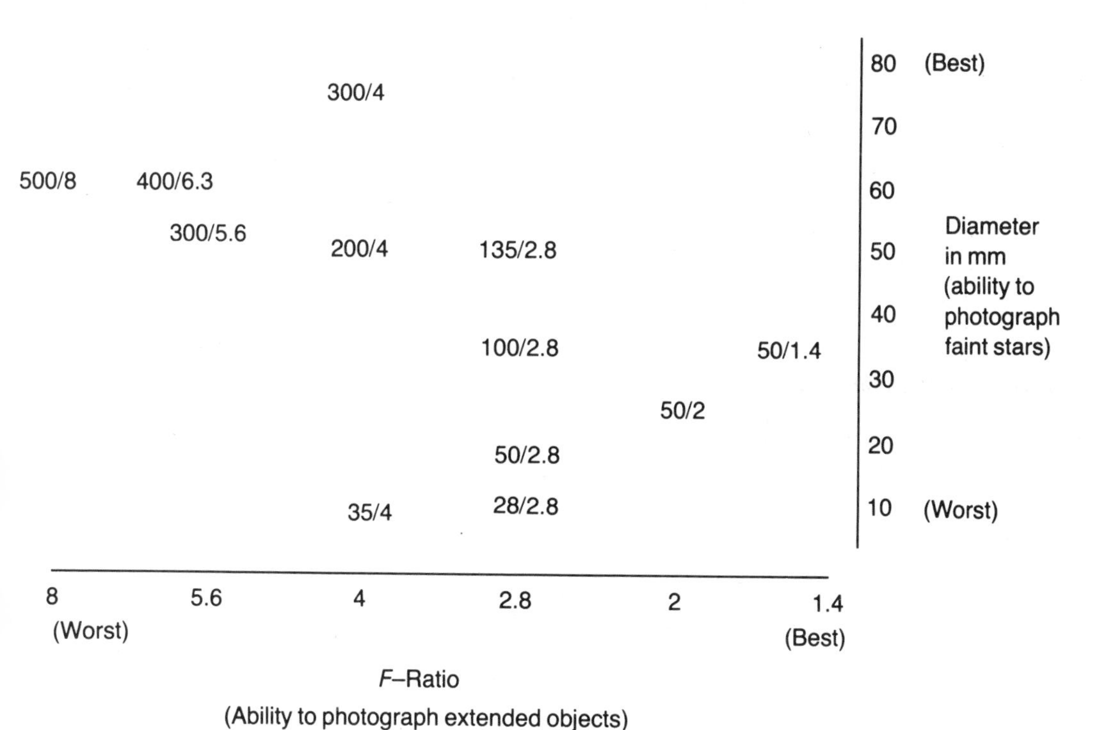

Table 7.4 also gives two other quantities that relate to the performance of the lens at the edge of the field. Because of off-axis aberrations, all lenses perform worse at the edges of the field than in the center; with a fast normal or wide-angle lens operating wide open, the star images in the corners of the picture are usually visibly blurred or distorted. This is a result of the way lenses are designed: the designer takes a given design and makes it faster and faster until its performance is almost, but not quite, unacceptable for ordinary (earthbound) photography. Since stars are more demanding subjects than human beings, a lens that is within acceptable limits on earth may be quite inadequate when aimed at the sky.

The remedy is, of course, to stop the lens down. Its maximum aperture was satisfactory by the designer's criteria but not by yours; your own maximum aperature will generally be one stop smaller (*f*/2 on an *f*/1.4 lens, for example). Such measures are not usually necessary for lenses longer than 100 mm or slower than *f*/2.8, or for certain lenses of very high quality.

Table 7.4. *Focal length, field of view, and edge-of-field effects for various focal lengths (on 35-mm film)*

Focal length (mm)	Field of view (degrees)	Relative scale enlarge-ment at corners (center = 1.00)	Per cent illumination in corners ($\cos^4$ law), assuming lens does not compensate
18	67.4 × 90.0	2.2	20
24	53.1 × 73.7	1.7	35
28	46.4 × 65.5	1.5	44
35	37.8 × 54.4	1.3	57
50	27.0 × 39.6	1.16	74
75	18.2 × 27.0	1.07	87
100	13.7 × 20.4	1.04	92
135	10.2 × 15.2	1.02	96
150	9.15 × 13.7	1.02	97
200	6.87 × 10.3	1.01	98
300	4.58 × 6.87	1.00	99
400	3.44 × 5.15	1.00	100
500	2.75 × 4.12	1.00	100

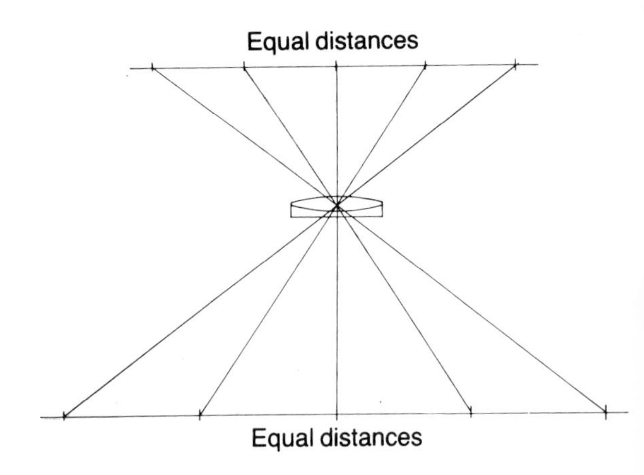

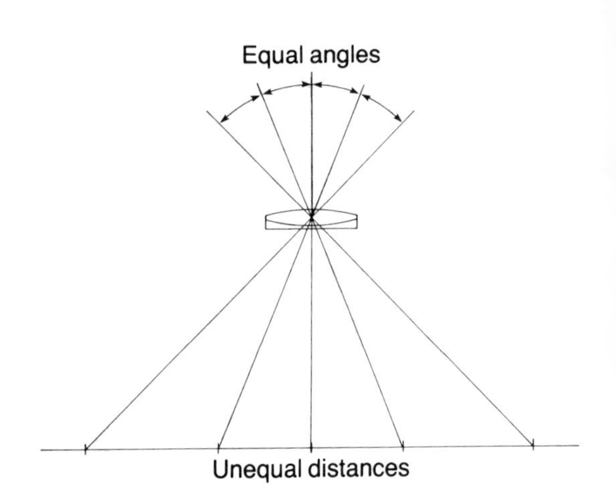

Fig. 7.6. A distortion-free lens reproduces equal distances as equal distances. Hence, by definition, it reproduces equal angles as unequal distances.

But even if you eliminate off-axis aberrations, the edges of the field will still show some peculiarities, at least with a wide-angle lens. This is where the additional numbers in Table 7.4 come in. First, as Fig. 7.6 shows, the angular scale will be larger at the edges than in the center. This is because you are projecting the sky – which is effectively a hollow sphere of infinite diameter viewed from the center – onto a flat film plane. (It's essentially an inside-out version of the old problem of how to make a flat map of the round earth.) If the lens will correctly reproduce a flat target on flat film, it is free of distortion; the change in angular scale is a purely geometrical phenomenon, and can be computed as follows:

$$\text{relative scale} = 1 / (\cos t)^2$$

where t is the distance off axis, expressed as an angle, or

$$\text{relative scale} = 1 / (\cos \arctan w/F)^2$$

where w is the distance from the center of the image on the film and F is the focal length, expressed in the same units. The third column of Table 7.4, then, gives the relative scale for a point 20 mm off-axis, in the corner of a 35-mm frame, and as you can see the effect is appreciable only for wide-angle lenses. Even then, it does not necessarily detract from the visual appeal of the pictures, though it is worth knowing about if you want to measure positions on photographs.

The other number, in the rightmost column, relates to the fall-off of light at the edge of the field – again, not necessarily a lens defect, but always to some extent an inevitable geometrical phenomenon.

Just as they tolerate some edge-of-field blurring, most lens makers tolerate some vignetting at the lens's widest aperture. Vignetting results from lens elements or stops that are not quite large enough to let all the light through; making the stops slightly too small is an excellent way to reduce reflections and lens flare. But the true vignetting, if any, should disappear when you stop the lens down. Any light fall-off that remains is the result of what is known as the $\cos^4$ law or cosine-four law:

$$\text{relative brightness} = (\cos t)^4$$

where, as before, t is the distance off axis, expressed as an angle.

Scale enlargement accounts for two of the four cosine factors in this formula. The other two come from the fact that the light at the edge of the field is not hitting the film square-on, and hence is spread over a larger area, and the fact that the back opening (exit pupil) of the lens does not face the film squarely, and hence has a smaller effective area.

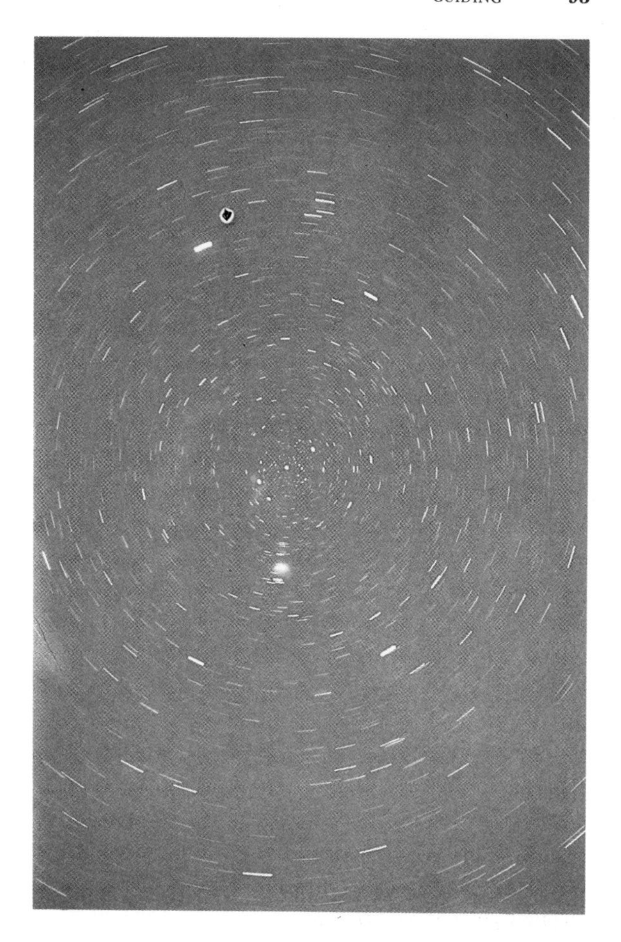

Fig. 7.7. *This is what happens if you make heroic guiding corrections with a mount that it not properly aligned on the pole, or an alt-azimuth mount (whose 'pole' is straight overhead). (Akira Fujii)*

The $\cos^4$ law is not absolute. Modern wide-angle lenses counteract it by making the entrance pupil larger off axis than on axis. You can observe this by looking at the lens from different angles. A few experimental ultra-wide-angle lenses for terrestrial photography have been made with absorbing filters that darken the center of the field to match the edges – hardly an approach that an astrophotographer would like. Fisheye lenses combat the problem in a different way, by introducing distortion to make the scale contract, rather than enlarge, at the edges.

Guiding

In order to take guided photographs of deep-sky objects, you need a telescope on an equatorial mount, a clock drive, and, preferably, an electronic drive corrector.

You may be able to do without a clock drive if you have some means of moving the telescope smoothly (a slow-motion knob or the like) and plenty of patience, and provided you are using a lens of relatively short focal length (say, 100 mm or less). Hand-guiding is much easier if you do it in periodic spurts rather than trying to achieve continuous motion. The trick is to determine your guiding tolerance (see below) and then aim the telescope so that the star you are guiding on is just ahead of the center of the field – ahead by half of the tolerance, in fact. Let it drift until it is behind by an equal amount, and then, with a motion that need not be particularly quick, put it back in its original position. Do this over and over until the exposure is complete.

This brings up an interesting question: if a clock drive is not absolutely necessary, what about an equatorial mount? Wouldn't an alt-azimuth mount with slow motions, or even a universal joint, work just as well, so long as you guide accurately?

The answer is no. If the telescope isn't actually rotating about an axis parallel to the earth's – as it does with an equatorial mount – you get rotation of the image; the guide star stays in the center of the field, but the rest of the picture rotates around it (Fig. 7.7).

To see why this is so, suppose you are photographing Orion using an alt-azimuth mount, guiding on the center star of Orion's belt. It is a crisp November evening, and Orion has just risen; his northwestern (upper right) shoulder is at the top of the picture. With incredible patience, you guide all night, until Orion is ready to set. At that point you notice that his other shoulder (the upper left one) is now highest and sets last. Since you have an alt-azimuth mount, the camera's 'up' and 'down' have remained the same as yours, and the guide star has remained in the middle, but the rest of Orion has rotated something like sixty degrees. As a result, everything except the guide star has left a curved trail.

What is more surprising to beginners is that, even if you have an excellent equatorial mount and clock drive, most of the time you still have to make guiding corrections by hand. There are several reasons why this is so:

1 Imperfect polar alignment (the equatorial mount is not perfectly parallel with the earth's axis).

2 Periodic errors in the gear train of the clock drive. (Most gear systems run alternately a bit fast, a bit slow, a bit fast again, and so on.)

3 Fluctuations in the frequency of the AC power supplied to the motor. (The exact speed of the motor in a clock drive depends on the frequency, not the voltage, of the alternating current. Although the *average* frequency of commercially supplied power is accurately regulated, moment-by-moment variations of a few per cent are common. A related problem is that a drive made with a standard electric clock motor makes one revolution in 24 hours, rather than the correct 23 hours and 56 minutes.)

4 Flexure of the telescope tube and mounting as the load shifts.

5 Atmospheric refraction (a change in the amount of air you are looking through slightly alters the apparent position of the object).

These five factors are listed in decreasing order with respect to your ability to do anything about them. In practice, a well-built telescope may track well enough to allow 5- or 10-minute exposures through a 50-mm lens without any guiding corrections; try it and see. If you get unsatisfactory results, first verify that the clock drive is indeed driving the telescope; is everything balanced? If there is a clutch adjustment to determine how tightly the clock drive is coupled to the telescope, try tightening it (but not by much; the telescope should still be able to move freely, with no risk of stripping gears).

The next thing to check is polar alignment. Sighting on Polaris is the usual method; here is a refinement of it, which will get you within something like a quarter of a degree of the true pole:

1 Make sure the declination circles really read 90° when the telescope is aimed parallel to its polar axis. Inaccuracy here is a major source of error. If the telescope and axis are truly parallel, then rotating the telescope about the polar axis will make stars seem to whirl around the center of the field rather than moving out of it.

2 Align roughly on Polaris. Then aim the telescope at a convenient star such as Capella (5^h16^m) or Mizar (13^h24^m) and set the right ascension circle to show the star's right ascension. In effect, you are using the star as a sidereal clock.

3 Aim the telescope, by setting circles, to the coordinates of Polaris (2^h26^m, +89.2°, epoch 1995). Then adjust the mount to get Polaris centered in the field.

For even greater accuracy you can do further checking as follows. Choose a star that is approximately overhead and try to track it for some time with a high-power eyepiece. If it seems to drift southward, the polar axis is too far west. Then choose a star that is about 20° above the eastern horizon and track it: if it drifts northward, the polar axis is aimed too high, and if it drifts southward, the polar axis is aimed too low. By repeating these tests and making successive corrections, you can achieve very accurate polar alignment.

Other sensitive, though time-consuming, tests are given by J. B. Sidgwick in Chapter 16 of *The Amateur Astronomer's Handbook* (New York: Dover, 1980). They are almost always unnecessary: if you can align within half a degree, the total error during a one-hour exposure of a star on the equator will amount to roughly 60 arc-seconds of drift and less

Table 7.5. Guiding tolerances for various focal lengths

Focal length (mm)	Guiding tolerance	Drift time in seconds (at equator)
18	290″ (=4.8′)	20
24	215″ (=3.6′)	14
28	185″ (=3.1′)	12
35	145″ (=2.4′)	10
50	105″ (=1.8′)	7
100	50″	3.4
135	40″	2.5
200	25″	1.7
300	17″	1.1
500	10″	0.7
800	6.5″	0.4
1000	5.2″	–
1250	4.1″	–
1500	3.4″	–
2000	2.6″	–
2500	2.1″	–

than a sixth of a degree of image rotation, or proportionally less for either a shorter exposure or a smaller alignment error.

The way to deal with the other sources of inaccuracy, especially the ever-present periodic gear error, is to make guiding corrections by hand. When photographing through lenses of up to about 200 mm focal length, you can often get away with checking the tracking every two or three minutes; with longer focal lengths, and especially when you are photographing through the telescope as described later in this chapter, guiding has to be continuous.

The normal technique is to use an eyepiece with crosshairs and center a slightly out-of-focus star image where the crosshairs intersect. A suitable eyepiece, which should be of about 6 to 12 mm focal length and need not be very good optically, can be salvaged out of an old finder or made by gluing crosshairs across the field stop of one of your less favorite eyepieces. (Spiderweb is ideal for making crosshairs, except that it is difficult to work with and easily damaged; most human hair is too thick, but a contribution from a baby or a person with thin blond hair will generally do. Thin hair-like filaments can be made out of some kinds of glue; also, you can make an excellent reticle, with crosshairs of any thickness, out of photographic film by photographing a suitable target.) Eyepieces with illuminated reticles are available commercially.

In order for the technique to work, you need to know how accurately you must guide. Assuming that you consider a resolution of 40 lines per millimeter on 35-mm film to be acceptably sharp, the formula is:

$$\text{guiding tolerance} = 2 \arctan (0.0125 / F)$$

where F is the effective focal length in millimeters. Table 7.5 shows the tolerances for a range of focal lengths.

Merely knowing the tolerance in arc-seconds is not enough: you must also know what a given number of arc-seconds looks like through your telescope. There are two ways to determine this. For relatively large tolerances, you can take advantage of the fact that, if the clock drive is not running, a star on or near the celestial equator appears to drift 15 arc-seconds in one second of time; if you want to see what a distance of 150″ looks like, simply center the telescope on a suitable star, turn off the drive, and wait 10 seconds. (Again, see Table 7.5 for equivalent drift times.) In the case of smaller tolerances, you can judge distances by observing double stars of known separation. Then record, for future reference, what your guiding tolerance looks like, compared to, say, the width of the crosshairs (for example, 'guiding tolerance for 300-mm lens = 4 crosshair widths').

You must guide both in declination and in right ascension. Guiding in declination is done with a mechanical slow motion. Right ascension guiding, on the other hand, is usually done by varying the speed of the clock drive – which in turn is accomplished by varying the frequency of the AC power supplied to it. Typically, you have a 'fast' button that increases the speed about 15%, a 'slow' button that decreases it by the same amount, and a control to adjust the speed at which the drive runs when neither of the buttons is being pressed. Most electronic guiders operate off a 12-volt DC supply (for example, a car battery); a power consumption of 0.8 ampere is typical. The output is a distorted square wave that causes the drive motor to make a louder than normal humming noise but otherwise does no harm; the voltage varies quite a bit, with no effect on the motor. Plans for building a guider of this type are given in Appendix D.

Two refinements are becoming common. One is to operate the declination slow motion with a small electric motor, so that you can do all your guiding from a single set of buttons or even a joystick-type control. The other, considerably more elaborate and just beginning to be available commercially, is to automate the guiding process entirely; the human observer is replaced by a photocell, and an analog computer circuit makes corrections as necessary.

Note that an electronic guider is exactly what you need if you want to operate an American telescope in Great Britain, or vice versa; use a guider that generates the voltage required by the telescope, which need not be the same as the local line (mains) voltage. But if you want to operate a telescope in the southern hemisphere, whose clock drive was designed for use north of the equator, you'll have to replace the motor with one that rotates in the reverse direction (which may not be difficult; clock drive motors are the same type as those used on appliance timers, and replacements are easy to obtain).

Films, filters, and fog

When you make a long exposure, you are relying on the film to accumulate faint light to make a usable image. But if the light is faint and the exposure is long, the film performs less efficiently than you might expect; not all of the light is captured. This phenomenon is known as *reciprocity failure*. As a rule of thumb, if you are using a film such as Tri-X Pan, then in order to double the amount of light captured in a long exposure, you must multiply the exposure time, not by two, but by three. (This corresponds to a Schwarzschild exponent of 0.63 as defined in Chapter 9. Some newer films are not as bad.)

Let's see what this means in practice. Suppose you have taken a 5-minute exposure and you want to take another that captures twice as much light (that is, shows a one-stop exposure increase). This requires an exposure of 15 minutes, not 10 minutes as you might expect. Make another one-stop increase, and you have a 45-minute exposure; another, and you're exposing for 2 hours and 15 minutes; another, and the exposure will take nearly 7 hours, if you can guide that long. If you could get rid of reciprocity failure, the equivalent of your 7-hour exposure would take you only an hour and 20 minutes.

The most effective way to eliminate reciprocity failure is to chill the emulsion. This is normally done in a *cold camera*, a special telescope-mounted device that holds the back of the film against a chamber full of dry ice. In order to prevent frost from forming on the emulsion, the film is insulated from the outside air by either a vacuum chamber or a thick plug of optical-quality plastic.

Cold cameras give better picture quality than other ways of dealing with reciprocity failure, since cooling the emulsion does not introduce fog or shift the color balance. But they are technically demanding; not only do you have to deal with dry ice makers and vacuum pumps, but you usually have to cut the film into small pieces and expose it one frame at a time. A cold camera is available commercially from Celestron International (see Appendix A); for plans for a home-built one, see *The Cambridge Deep-Sky Album* by Jack Newton and

Philip Teece (Cambridge, England: Cambridge University Press, 1983). Plate 7.6 shows two outstanding examples of amateur cold-camera photography.

Incidentally, there is evidence that the temperatures used in amateur cold cameras may be too low. The best control of reciprocity failure occurs around −20 ° or −30 °C (*Kodak Plates and Films for Scientific Photography*, 1973, p. 22), whereas the temperature of dry ice is −43 °C, low enough to produce an undesirable reduction in film speed. Even 0 °C (32 °F) is markedly better than room temperature – which suggests that the clear air is not the only reason our winter star-field pictures turn out so well.

Another way to reduce reciprocity failure, and at the same time pick up more faint stars even in short exposures, is to *preflash* the film, exposing it to a small amount of light in advance of the exposure. The idea here is that faint light has more of an effect when it adds to an already existing fog level than when the film's response to it has to start from zero.

Changes in environment can also reduce reciprocity failure. Simply storing your film in a vacuum for a few hours or days before exposure can improve its performance. So can baking it, in a vacuum or in air, at 65 °C (150 °F) for a couple of hours, though a considerable amount of fog results.

But the best kind of pre-treatment consists of baking the film in hydrogen gas, or in a mixture of gases containing hydrogen. This is the 'gas hypersensitization' referred to in the literature, and a non-flammable mixture of nitrogen and hydrogen called 'forming gas' is normally used. A typical procedure is to place the film in a vacuum for a few minutes to get rid of moisture, then bake it in 92% nitrogen, 8% hydrogen, for three days at 30 °C (86 °F), or for a shorter time at a higher temperature. For more information, see Roger Sliva, 'Hypersensitizing' (*Astronomy*, April 1981, pp. 39–42, and May 1981, pp. 48–50), or write to Lumicon, Inc. (see Appendix A), which sells the necessary equipment and materials. Gas-hypersensitized film is available commercially from Lumicon and other suppliers.

CAUTION

Hydrogen gas is flammable and explosive (even more so than natural gas or cooking gas). In addition, all compressed gases pose the risk that apparatus will burst under high pressure. Use care and get expert advice when experimenting with gas hypersensitization.

Gas hypersensitization virtually eliminates reciprocity failure and introduces only a small amount of fog, especially if the film is stored in a freezer after treatment. Freezing the film also reduces the tendency of the speed increase to fade away with time. Technical Pan 2415 shows a dramatic speed increase when hypersensitized and can be stored at room temperature for days, or in a freezer for weeks, without adverse effects. Color slide films have more stringent storage requirements but are still usable after two or three days out of the freezer, just long enough to get the film from the supplier; Fujichrome 100 gives particularly good results.

Some newer films suffer relatively little reciprocity failure even without hypersensitizing. This is the case with the Kodak T-Max series (especially T-Max 100) and with Fujichrome 100 (which, however, suffers a strong color shift towards green). See Chapter 9 for more information.

Once you get rid of reciprocity failure, you quickly run into sky fog – that is, you find that even in a relatively short exposure, the film records enough sky glow to cover up some star images (Fig. 7.8). Sky fog sets an absolute limit to the magnitude of the faintest stars you can photograph

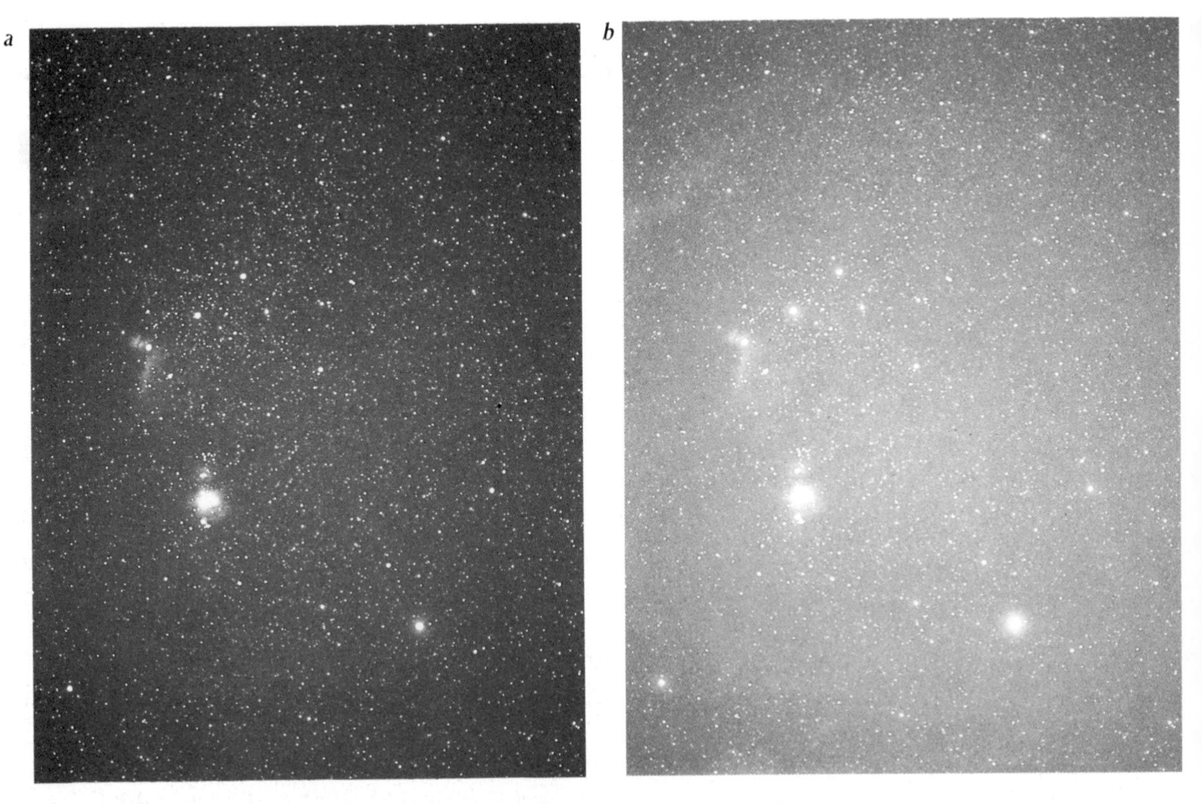

Fig. 7.8. *Illustrating sky fog problems. Two photographs of southern Orion with a 100-mm lens at f/2.8 on Fujichrome 400: (a) 10 minutes, (b) 24 minutes. The shorter exposure shows more stars. (By the author)*

with a given lens, regardless of exposure time.

The only way to reduce sky fog, apart from moving to a remote mountaintop, is to photograph through a filter in the hope of blocking out the parts of the spectrum at which the sky fog is worst. A yellow filter usually helps, but a better technique is to photograph in the extreme red, using a #25 or #29 deep red filter with Technical Pan film. The fog-penetrating ability of such a combination is impressive (Fig. 7.9), and it is worth trying even in environments that you would normally consider too urban for deep-sky work. Emission nebulae record best because they emit most of their light in just the part of the spectrum that the #29 filter passes. Special band-blocking 'nebulae' and 'galaxy' filters designed for visual use are also worth trying; so are magenta filters, even with color film.

Ordinary black-and-white films (including T-Max) do *not* respond to the deep red light emitted by ionized hydrogen and therefore do not photograph nebulae well. Here Technical Pan and colour films have the advantage.

Fig. 7.9. *A suite of deep-sky photographs by Akira Fujii; all are 30-minute exposures on gas-hypersensitized Technical Pan 2415 with a Canon 300-mm telephoto lens at f/2.8 and a red filter. (a) The Orion Nebula and Horsehead Nebula; and, overleaf (b) the North America Nebula; (c) the 'California' Nebula in Perseus; (d) the Double Cluster in Perseus.*

b

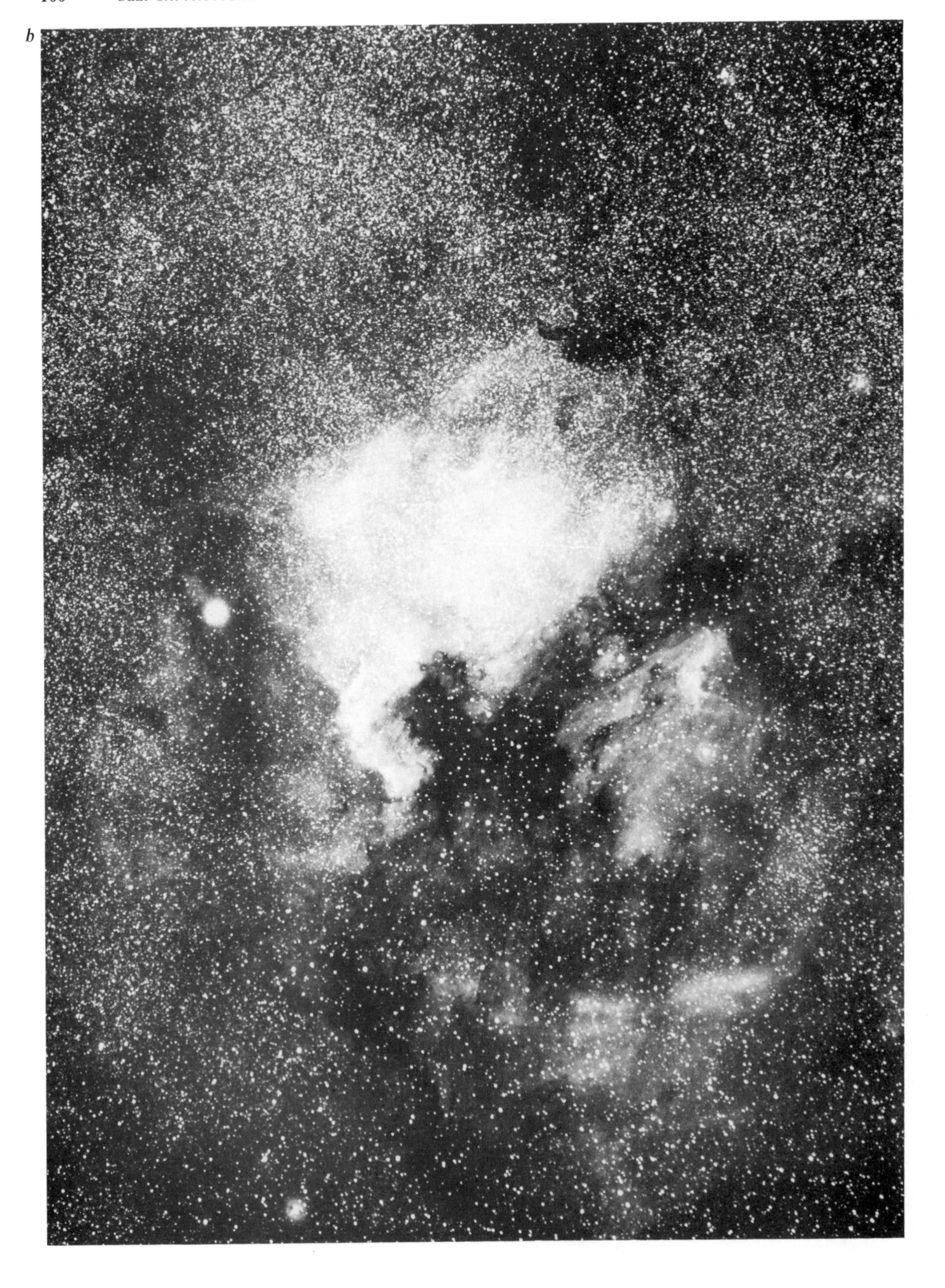

c

d

Through the telescope

The advent of cold cameras and hypersensitized film has made it possible for amateurs to take impressive photographs of spiral galaxies, globular clusters and similar objects, often rivaling the results large observatories were producing only a few decades ago. The required optical configurations, prime focus and compression, were discussed in Chapter 5. The main challenge is guiding – how do you guide if the main telescope is occupied with taking a picture?

One obvious solution is to use a *guidescope* – a second telescope, similar in focal length to the main one, mounted piggy-back on the instrument with which you are taking the picture. The guidescope should have a long enough focal length, and operate at a high enough magnification, to make it easy to stay within the guiding tolerance of the main instrument. It does not have to be large or have particularly good optics; a 5-cm (2-inch) refractor with a strong Barlow lens is typical, and folded refractors or other uncommon configurations can be used for compactness.

Nor does the guidescope have to be aimed at the same object as the main telescope; any bright star in the vicinity will do. (A guide star more than a few degrees away, however, could introduce errors because of atmospheric refraction.)

Guidescopes do present the problem of flexure. If, due to shifting weight, the guidescope bends as much as 1/500 of a degree relative to the main telescope, you can go outside of the guiding tolerance without knowing it. Flexure is seldom a problem with relatively short exposures (30 minutes or less); if you have problems with it, don't overlook the possibility of a shifting optical component, such as a diagonal mirror. For procedures to detect and eliminate flexure, see Robert Provin and Brad Wallis, 'Guiding', *Astronomy*, December 1980, pp. 39–42.

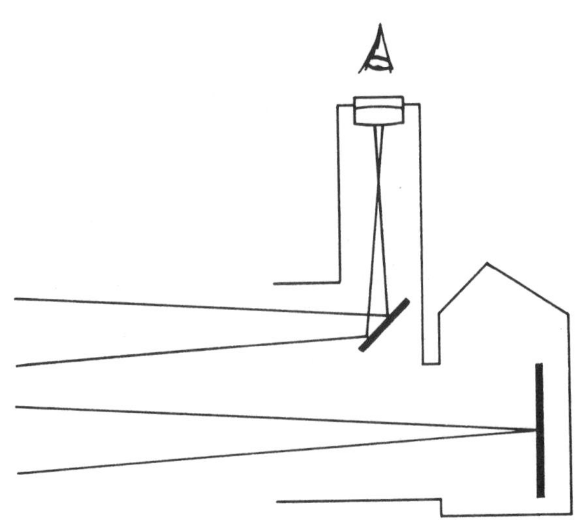

Fig. 7.10. *The principle of an off-axis guider. In practice, the system feeding the eyepiece usually includes a Barlow lens, both to give a larger image and to allow a longer tube.*

To get rid of flexure problems almost entirely, you can use an *off-axis guider* (Figs. 7.10 and 7.11). This device intercepts a small, outlying area of the image formed by the main telescope and allows you to view it through a guiding eyepiece. The only problem is that you almost always have to guide on a rather faint star, since you cannot see the main object being photographed.

Apart from a large telescope, the ultimate instrument for deep-sky photography is probably the *Schmidt camera*, an instrument built like a Schmidt–Cassegrain telescope except that the film itself is mounted in place of the secondary mirror. The result is a very low *f*-ratio with excellent sharpness over a wide field. Celestron manufactures 14-cm (5.5-inch) *f*/1.65 and 20-cm (8-inch) *f*/1.5 Schmidt cameras designed to be piggy-backed on larger telescopes. The pictures in Plate 7.1 are examples of what these instruments can do.

a

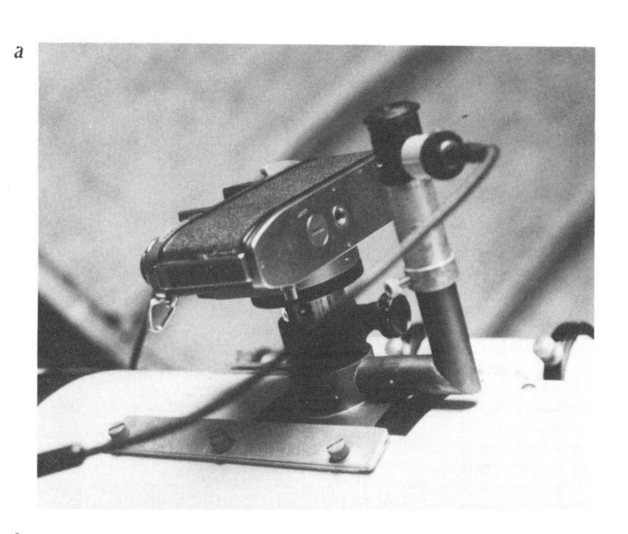

Fig. 7.11. *(a) The off-axis guider on Walter A. Singer's homebuilt 20-cm (8-inch) f/4 Newtonian.*
(b and c) Pictures taken with it: the Pleiades (15 minutes) and the Lagoon Nebula (20 minutes) on pushed Fujichrome.

b c

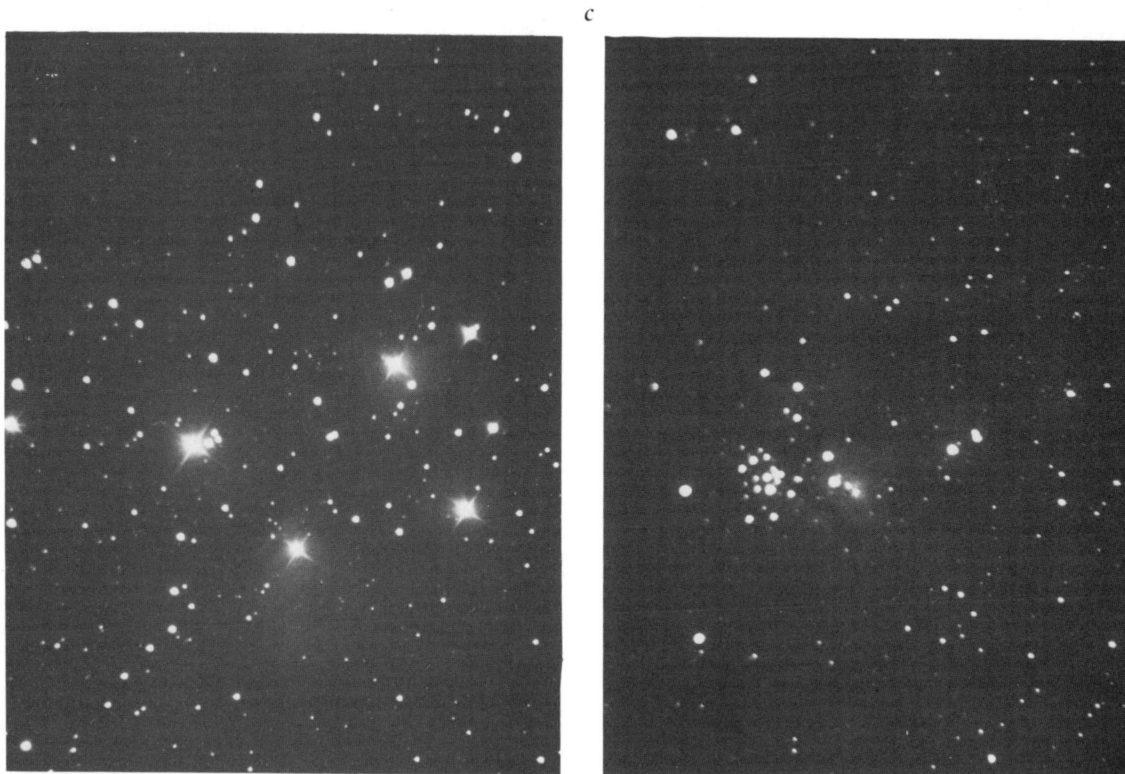

Keeping warm while observing

Astrophotography is one of the few human activities that involve standing perfectly still outdoors at night in the middle of the winter. (Visual astronomers, with no continuous guiding to do, can at least walk around occasionally.) As a result, the astrophotographer is exposed to colder weather than anyone else living in the same part of the world – often, in fact, colder weather than his or her neighbors will even believe occurs there. The necessary clothing and the knowledge of how to wear it are seldom available locally, especially in warm climates. Growing up on the Georgia–Florida border, I had to learn the art of dressing for cold weather the hard way; what follow are a few practical observations.

People from warm climates are often tempted to sink all their resources into one enormous coat that looks as though it ought to be warm simply by virtue of its bulk. Resist this temptation. Overcoats are meant to keep out rain and snow, not cold air. They leave large areas of your body (the head, neck and lower legs) uncovered, they impede your movements, and they are often heavy enough to be uncomfortable after a couple of hours.

Instead, dress in layers in order to trap warm air close to your body. Wear a sweater under your jacket. Consider using two or three layers in place of what is ordinarily just a shirt. Keep your legs warm with long underwear, possibly two pairs at once. (Many garment manufacturers seem to have forgotten that the purpose of long underwear is to warm the legs, not to compress them; I get much better results from loose-fitting cotton and wool than from elastic mesh.) You can even wear flannel pajamas under your clothes for extra warmth.

Cold feet often result from cold legs and are best combatted with long underwear. Thick wool socks, possibly with thin cotton socks inside them, are a great help. Shoes should have thick rubber soles, rather than thin leather ones, in order to reduce heat conduction between your feet and the cold, damp ground; putting down a rubber mat or a scrap of carpet where you are standing can help considerably.

A knitted sock hat is obligatory below about 7 °C (45 °F); at much lower temperatures a fur hat or Russian calpac is better. (A hat with a brim would bump into the telescope.) Fill the gap between collar and hat with a thick knitted scarf. Gloves are helpful but should be fairly thin if you are to manipulate controls. Catalytic (fuel-burning) hand warmers can be useful not only in keeping your fingers limber but also in getting dew off lenses.

The best suppliers of cold-weather clothing are camping and hunting outfitters; if the pickings are slim in your area, write to L. L. Bean and Company (Freeport, Maine 04033, USA), which sells a full line of high-quality products by mail. If the price seems high, consider it as part of your equipment budget; anything that doubles the number of nights a year that you can photograph is effectively doubling the value of your telescope and camera.

Part III
EQUIPMENT AND MATERIALS

8
Cameras, lenses, and telescopes

The 35-mm SLR

The 35-mm single-lens reflex is probably the single most useful type of camera for amateur astrophotography. It is small and light enough to be attached to a telescope without overbalancing it; it combines the versatility of ground-glass focusing with precision rivaling that of rangefinder cameras; and its film format is about the same size as the eyepiece tubes of amateur telescopes, so that the restricted image size does not waste much film. Moreover, a wider variety of emulsions is available on 35-mm film than in any other format except for sheet film.

A single-lens reflex (SLR) is a camera in which viewing, focusing, and picture taking are done through the same lens (Fig. 8.1). This is accomplished by means of a mirror that intercepts the incoming light from the lens and deflects it to a ground glass screen that can be viewed through a prism and magnifying lens. When a picture is taken, the mirror flips up, out of the way; the shutter,

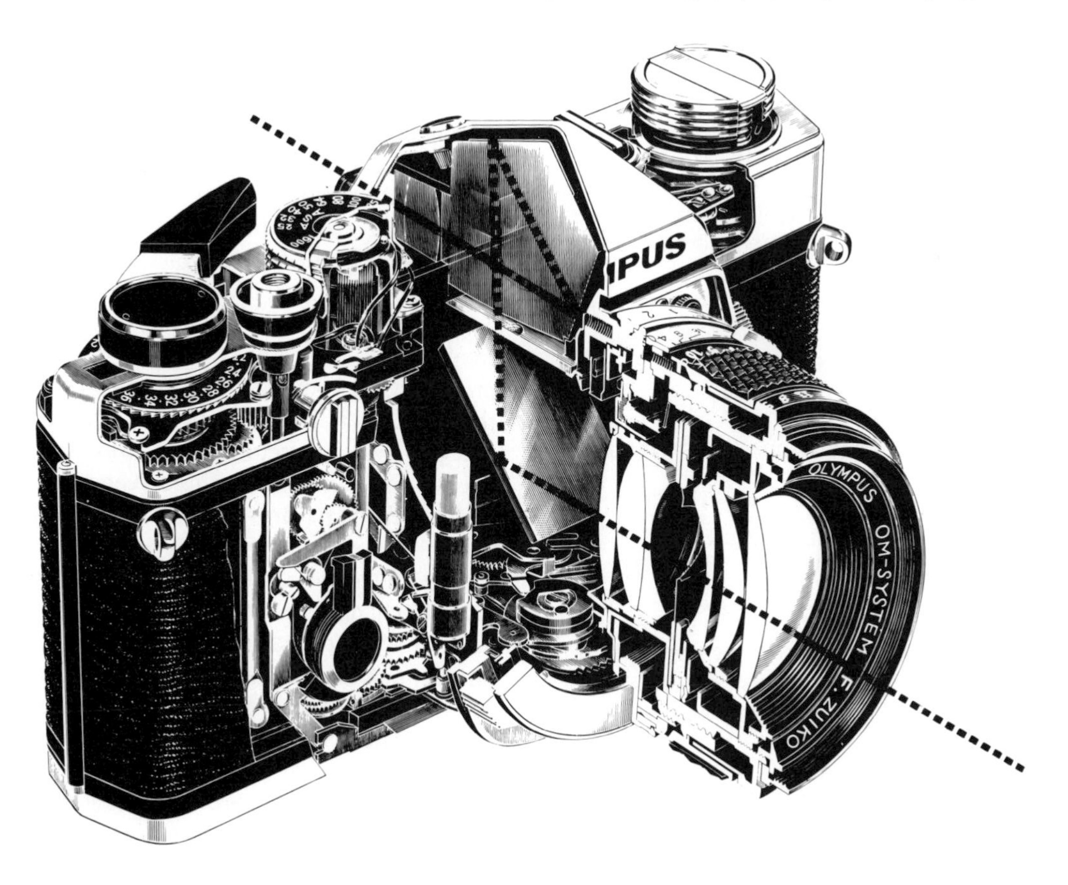

Fig. 8.1. *Cross section of a 35-mm SLR, showing the light path for viewing. (Olympus Optical Company)*

located behind the mirror, then opens and closes, and the mirror returns to its original position.

Some SLRs are, of course, more suitable for astronomical use than others (Table 8.1). As noted in Chapter 6, it is very useful to have interchangeable focusing screens; a focusing screen that cannot be changed should include a fine matte area in or near the center, typically in a broad ring around the central microprism or split-image area. Another desirable feature is a button that allows you to reduce vibration by locking the camera's mirror in the 'up' position prior to taking a picture.

Table 8.1. *Choosing a 35-mm SLR for astrophotography: features to look for*

Essential or almost essential	Reflex (SLR) focusing Interchangeable lenses, preferably with the ability to use T-adapters (this rules out certain leaf-shutter SLRs) Ability to make time exposures (preferably without running down batteries)
Desirable	Mirror lock Interchangeable focusing screens
Unnecessary	Autofocus Automatic exposure Light meter
Acceptable defects in a used camera	Failed light meter Inaccurate or partly inoperative shutter (if 'B' still works and there is no risk of jamming)
Defects to watch out for	Inaccurate focusing (focusing screen not matching film plane) Shutter prone to jamming

There are some features that you do not need; among them are automatic focusing, automatic exposure, or indeed any kind of light meter. In fact, some automatic-exposure cameras exhaust their batteries during time exposures; check into this when choosing a camera, though if you need the automatic-exposure feature for terrestrial photography, it may be cheaper to keep a generous supply of batteries or rig up an external power supply than to buy a second camera.

If you want a camera solely for astronomical work, you can save money by buying a used camera that is not completely operational. The single most common SLR breakdown is failure of the light meter, which of course is not needed for astronomical photography. Shutter breakdowns are also very common and may not be fatal from the astrophotographer's viewpoint; most shutters use separate mechanisms for high- and low-speed ranges (above and below 1/30 or the like), and it is common for one range to fail while the other continues to operate – or for the failure to consist of a simple, though dramatic, loss of accuracy. The result is a camera that is useless for ordinary photography and can be bought for a small fraction of its original price, though from the astronomical point of view there's hardly anything wrong with it. As long as 'B' (the time-exposure setting) still works, the shutter does not jam, and the focusing screen accurately matches the film plane, a junked camera may prove ideal.

The single most vital feature of an SLR is that it has interchangeable lenses and can be coupled to any type of optical system. There are many different kinds of lens mounts, and, in general, adapters to fit one camera's lenses onto another camera are seldom feasible. The space taken up by an adapter would leave the lens too far from the film to focus on infinity. The exceptions are when the second camera body is thinner than the first (e.g. Nikon lenses on

Olympus) or when the adapter contains a teleconverter. For more information on adapters, consult a large, well-stocked camera dealer.

Two systems for standardizing lens mounts appeared in the 1960s, the universal (Pentax–Praktica) screw mount and the T-system. The universal screw mount is 42 mm in diameter and has threads approximately a millimeter apart; in its heyday, it was used by dozens of manufacturers, though even Pentax has now abandoned it. Both used and new screw-mount lenses, however, are still in abundant supply at the moment.

The T-system, introduced by Spiratone in New York, is still used extensively for special-purpose lenses and camera accessories. Like a universal screw mount, a T-mount is 42 mm in diameter, but it has threads about 0.6 mm apart and is designed to fit, not directly into a camera, but into a *T-adapter (T-ring)* which, in turn, fits whatever type of lens mount the camera is designed to accept. Most commercially made telescope-to-camera adapters require a T-ring. In addition, you can get *T-flanges* that perform the opposite function and adapt a camera lens to fit into a female T-mount; a T-flange and T-ring together make a short extension tube and can be used to adapt lenses or other accessories from one camera to another if the increased lens-to-film distance is not a problem.

Both the T-system and the universal screw mount are on the decline partly because most photographers prefer the faster lens-changing that is possible with bayonet mounts, and partly because of the additional functions that modern lens mounts are called on to perform. Besides holding the lens on to the camera in the correct position, a modern lens mount must provide for automatic diaphragm operation – that is, it must allow the user to view and focus with the lens wide open, and then, under camera control, close down to the selected *f*-stop when the picture is taken. The universal screw mount does this by means of an optional pin at the bottom of the lens mount; the T-system does not provide for it at all. Moreover, most present-day SLRs allow the user to make meter readings with the lens wide open; an extra pin or lever transmits to the camera an indication of the *f*-stop to which the lens is set. Neither the universal screw mount nor the T-system allows this kind of metering.

Other types of cameras

The 35-mm SLR is of course not the only kind of camera that can be put to good use in astrophotography. Its most obvious rivals are 35-mm cameras that focus in some way other than through the lens. Such cameras can be used for practically the whole range of astrophotographic subjects, though the techniques available are limited to piggy-back photography with the camera's non-removable lens and afocal photography through the telescope. This is in itself enough to open up most of the realm of lunar, planetary, and wide-field photography. Much of my own early work was done with a Petri 7 rangefinder camera and a Voigtländer Vito B that had only a distance scale for focusing but whose 50-mm *f*/3.5 lens was remarkably sharp.

The other miniature formats, 126 (Kodak Instamatic) and 110 (Kodak Pocket Instamatic), are now nearly obsolete; only a few kinds of film are still available in these sizes. The image area of 126 film is about equivalent to 35-mm, but 110 film is considerably smaller; 110 users can expect roughly twice as much grain as on enlargements of the same size made from 35-mm film – which may not be too much, especially on Kodak Verichrome Pan and Kodacolor Gold 200, both of which are quite fine-grained. The shape of its characteristic curve suggests that Verichrome Pan would lend itself well to 'pushing' (producing an effective speed increase by overdevelopment), though I do not know how well this works in practice.

Many 6-cm (2¼-inch, 120/220/620) SLRs, such as the Bronica and Hasselblad, are just like their 35-mm counterparts, only larger. Except for the fact that the camera is heavier and requires more support and counterbalancing, the same principles apply. Twin-lens reflexes, though more limited in scope, are especially suitable for afocal photography because you can focus through the viewing lens, then shift the camera and make the exposure through the taking lens.

In planetary photography, there is no advantage to using large film formats, since the usual image size, determined by limitations of *f*-ratio, does not even fill the 35-mm field, much less anything larger.

In star-field photography, however, an interesting thing happens as you go from a smaller format to a larger one. Consider a 35-mm camera with a 50-mm normal lens and a 4×5-inch sheet-film camera with a 200-mm lens, both operating at *f*/3.5. The two lenses cover the same field of view, and they record extended objects equally well, since their *f*-ratios are the same. But their diameters are different – the larger camera has four times the clear aperture of the smaller one. This gives it, theoretically, a 3-magnitude advantage in picking up faint stars. Not all of the advantage is realized in practice, since if the two negatives are enlarged to the same size, some of the star images from the 4×5 film will come out too small to see. But it is clear that larger formats are better for producing pictures rich with stars.

Should you, then, opt for as large a format as possible, limited only by your telescope's ability to support the camera and your ability to pay for the film? No. The situation is complicated by the fact that fast lenses are not available for large formats. You can get an *f*/1.2 normal lens for a 35-mm camera; the fastest lenses for a 6-cm-format camera are about *f*/2.8, and for a view camera, about *f*/4. Hence, although the 4×5 camera in our example has the edge when photographing stars, a 35-mm camera with a fast lens can easily surpass it in photographing nebulosity. With Kodak Technical Pan film, the 35-mm picture will hardly show any more grain than the larger one.

If you happen to have a good large-format lens, such as an old aerial photography lens, you may want to build a box-like sheet-film camera for it. Such a camera need not even have a ground glass, since, as long as you know the approximate image plane, you can focus by trial and error. Make a long exposure of a bright star passing through the center of the field; divide the exposure into 1- or 2-minute segments, moving the lens to a different position for each one, making marks and keeping records as you do so. In order to know which segment is the beginning and which is the end, make the last segment extra long. The developed negative will show you which of the five or ten settings you tried was best; you can then make another series in which the changes you make are smaller, until finally you achieve perfect focus and lock the lens permanently in place. A lenscap and/or the filmholder slide can take the place of a shutter.

b

Fig. 8.2. *Resolution is not the same as apparent sharpness. Picture (b) resolves more lines per millimeter (note the sharp corners), but picture (a) looks subjectively sharper because there is less smearing of light areas into dark.*

Lenses

The basic characteristics of lenses have already been dealt with in the foregoing chapters: focal length, *f*-ratio, and image scale in Chapter 5, and edge-of-field performance and the $\cos^4$ law in Chapter 7. The purpose of this section is to deal with lenses from a more practical standpoint.

After focal length and *f*-ratio, the lens specification that you hear the most about is resolving power, generally measured in lines per millimeter (lpm) and determined by photographing a special test target on extremely fine-grained film in a vibration-free environment. In general, a resolving power of 40 lines per millimeter or better (with 30 lpm at the edge of the field) is adequate for 35-mm photography and represents a minimum standard that reputable lenses seldom fall below. Higher quality lenses have resolutions more like 60 lpm (45 at edge), and a really excellent lens such as the 50-mm *f*/1.8 Nikkor or Olympus Zuiko can reach 80 or 90 lpm at the center of the field at *f*/5.6. (These figures are from test results published in *Modern Photography* magazine; they are typical of present-day lens technology.)

Lenses are always sharpest in the middle of their *f*-stop range. When a lens is operated wide open, uncorrected aberrations and manufacturing tolerances take their toll; at the smallest apertures, resolution is limited by diffraction (at *f*/16, the Rayleigh limit is 125 lpm, and the practical diffraction limit is about 60 lpm). Also, naturally enough, lens designers plan their aberration corrections to work best at the apertures they think will be used most frequently, around *f*/5.6 to *f*/11. In astronomy, it is usually the performance of the lens wide open, or nearly wide open, that matters – and hence the comparative ratings of various lenses are different than for terrestrial photography.

Moreover, as Fig. 8.2 illustrates, the resolving power of a lens is not the only thing that determines how sharp a picture looks. The presence or absence of reflections within the lens will also affect apparent sharpness. To see why, imagine a perfectly focused picture with a misty, smeared image superimposed on it. The misty, smeared effect does not reduce the resolving power at all – it does not remove any of the test target detail that forms the basis for resolution measurements – but it can certainly make the picture look blurred to the human eye. Remove all the mistiness and the picture will look sharper even though the actual resolution, in lines per millimeter, has not improved. For this reason, test reports often evaluate the contrast as well as the resolving power of a lens.

There are several additional points to consider, especially when you want to use lenses not originally designed for your type of camera. One is the size of film that the lens was originally designed to cover. A normal lens for a view camera and a telephoto lens for a miniature camera may both have a focal length of 200 mm – but they are not the same lens. The view-camera lens is, of course, designed to cover a much wider area on the film without vignetting. It is likely to have inferior resolving power, since view-camera negatives are seldom enlarged more than ×3

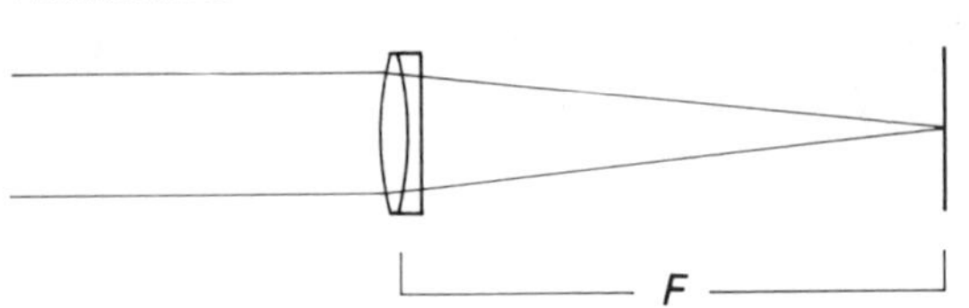

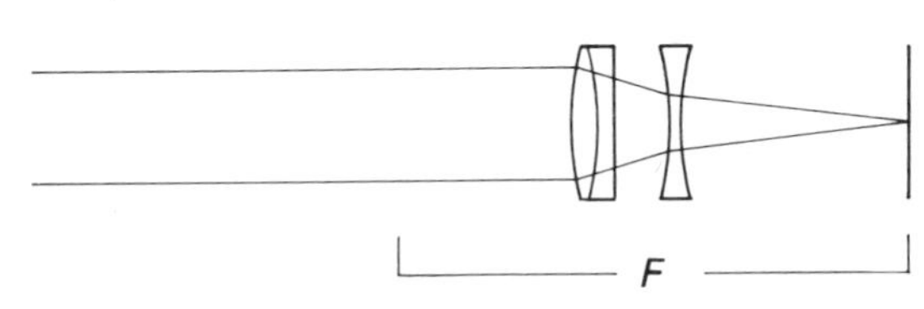

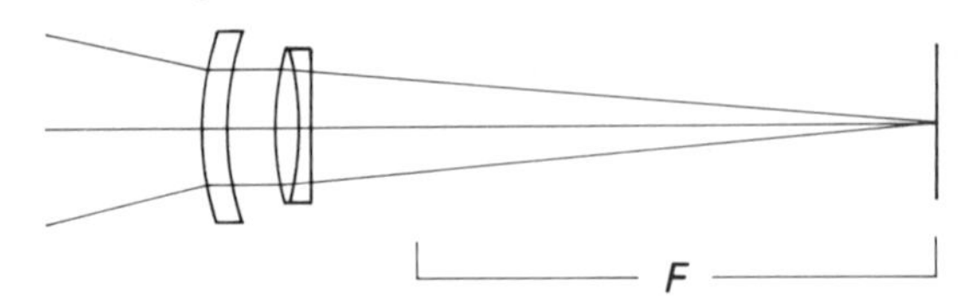

Fig. 8.3. *A telephoto lens is physically shorter than its focal length; a wide-angle lens is physically longer than its focal length. This makes it possible to mount lenses of widely varying focal lengths on the same camera without a bellows.*

or ×4, and it will certainly operate at a greater distance from the film.

Fig. 8.3 indicates why this is so. A telephoto lens is not simply a lens with a long focal length – it is a lens with a long *effective* focal length, designed to cover only a narrow field of view and to operate much closer to the film (hence taking up less space) than its focal length would suggest. Similarly, a wide-angle lens is not just a lens with a short focal length; it has to cover a wide angular field and operate at a lens-to-film distance greater than its focal length. Most telephoto lenses are actually negative projection systems; wide-angle lenses are negative projection systems turned backward. A zoom lens is a lens whose effective focal length can be changed by varying the spacing between elements.

Another factor is the reproduction ratio for which the lens was designed – that is, the ratio between the object-to-lens and lens-to-film distances. Macro lenses, copy lenses, and enlarging lenses are not suitable for astrophotography because their aberrations are corrected for reproduction ratios near one to one, whereas the ratio for an astronomical photograph is one to infinity. Moreover, copy and enlarging lenses are specially corrected to form a flat image of a flat object, as shown in Fig. 8.4, and this impairs their performance in astronomical applications. Projector lenses are even worse, since they are designed for speed and low cost, not resolving power.

If you use surplus aerial camera lenses – which, incidentally, can give good results with larger film formats – bear in mind that many such lenses were designed for use with yellow filters and suffer marked chromatic aberration when unfiltered white light goes through them. In fact, a #12 or #15 yellow filter can improve the image quality with ordinary telephoto lenses as well.

Among commercially available lenses for 35-mm cameras, the most expensive ones are not always the best for astronomical use, since some of the features that raise the price of a lens are irrelevant for astronomical work. Lenses for terrestrial use almost always have automatic diaphragms; the older-style preset or manual lenses are just as good for astronomical work, if you can find them. Further, lenses that are compact, light, and mechanically rugged sell for higher prices than bulkier, less solidly built lenses of equal optical quality – a difference well worth paying for if you are a photojournalist or travel photographer, but of little concern to an astrophotographer whose equipment will never be carried around the neck or treated roughly. Finally, zoom lenses are relatively unsuitable because, although they cost more than conventional lenses, they are neither as fast nor as sharp.

In general, lenses from major camera manufacturers (Nikon, Canon, Olympus, Leitz, and the like) are of higher optical quality than lower-

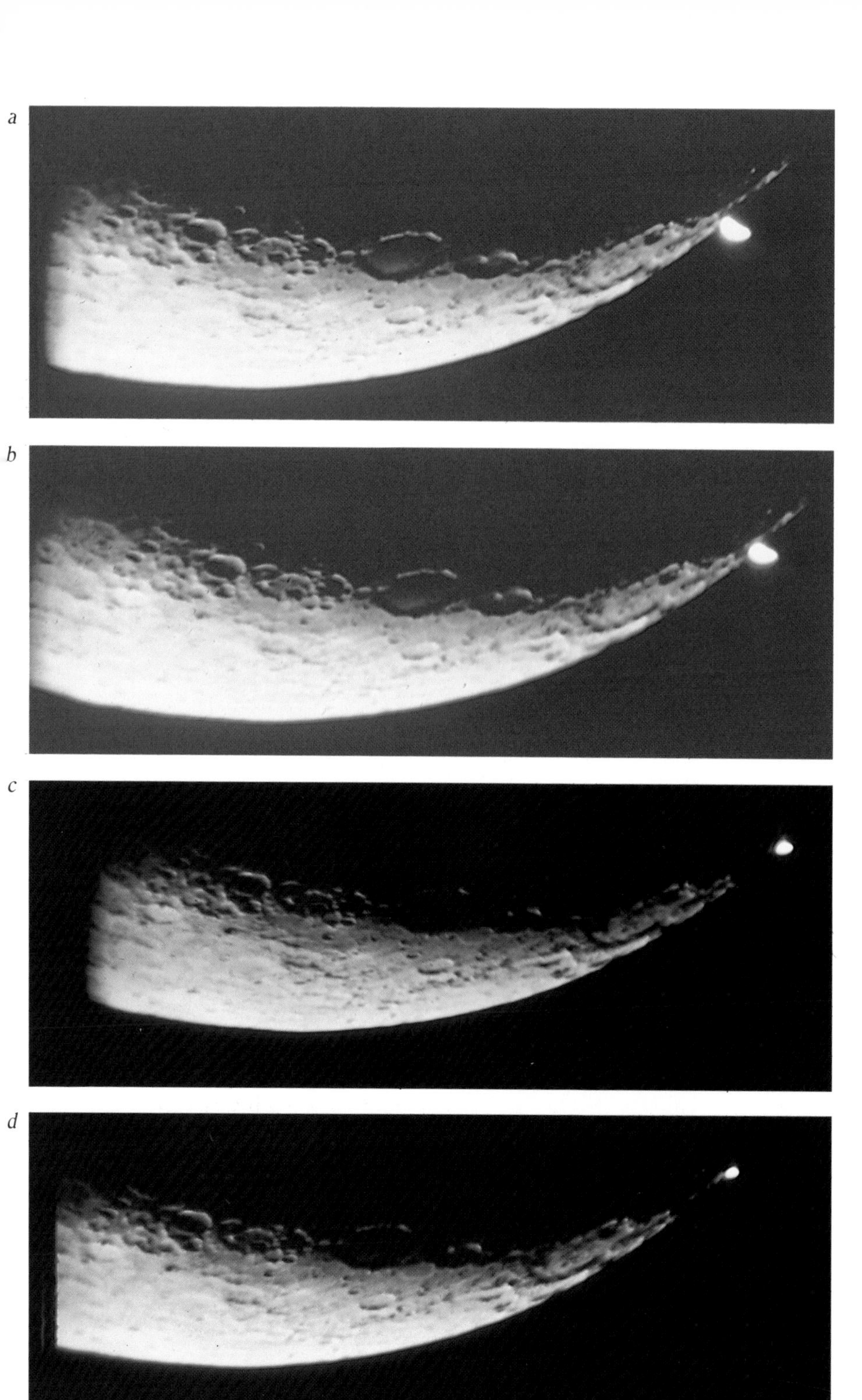

Plate 6.1. *Venus gradually disappears behind the edge of the moon in this series taken during the occultation of 26 December 1978 by Dale Lightfoot. Taken with a 20-cm (8-inch) f/5 Newtonian, positive projection with a 9-mm eyepiece to give an effective focal length of about 6500 mm (f/32); 1.5 to 2 seconds on Kodachrome 64.*

Plate 7.1. *Three types of deep-sky objects as photographed by George East: (a) the North America Nebula in Cygnus, a gas cloud so named because of its shape; (b) the dense star cluster M11 in Scutum; (c) the galaxy M31 in Andromeda. All are 15-minute exposures on Fujichrome 100 with a 20-cm (8-inch) f/1.5 Schmidt camera.*

a

b

c

Plate 7.2. *The Lagoon and Trifid nebulae in Sagittarius. Note the contrast between fluorescent gas (red) and dust which merely reflects starlight (blue). Fifteen minutes on Fujichrome 100 with a 20-cm (8-inch) f/1.5 Schmidt camera. (George East)*

Plate 7.3. *A wide-field view of Orion and Taurus through a 24-mm lens at f/2.8. Ten minutes on Fujichrome 400; the ambient temperature of −12 °C (10 °F) reduced reciprocity failure. Note the Rosette Nebula at the left. The North Orion Bubble is visible as a slight reddening of the sky background. (By the author)*

Plate 7.4. *A 10-minute exposure of southern Orion through a 100-mm lens at f/2.8 under the same conditions as Plate 7.3. The original slide was duplicated onto Kodachrome 64 to increase contrast. (By the author)*

Plate 7.5. *A hand-guided exposure of the summer Milky Way by Douglas Downing. Seven minutes on Fujichrome 100 with a 50-mm lens at f/1.9. Note the Lagoon Nebula (M8) at the lower right.*

a

Plate 7.6. *Two deep-sky photographs by Akira Fujii, both 50-minute exposures at the prime focus of a 30.5-cm (12-inch) f/5 reflector on Fujichrome 100 chilled with dry ice in a home-made cold camera. (a) The Orion Nebula; (b) the galaxy M51 in Canes Venatici, 10 million light years away.*

a

b

Plate 10.1. *A dramatic demonstration of what a slide duplicator can do. (a) The field of M8, 15 minutes on Ektachrome 200 with a 100-mm f/2.8 lens. (b) A slightly enlarged duplicate onto Kodachrome 25, made with light from a blue photoflood and exposed two stops more than the meter indicated. (c) An enlarged duplicate onto Ektachrome 200, exposed similarly. (Michael and Melody Covington)*

Plate 10.2. *The color picture that results from combining the three black-and-white exposures in Fig. 10.5. (Jim Baumgardt)*

a

b

Plate 10.3. *Two more color pictures formed by combining black-and-white exposures through red, green and blue filters. (a) The Pleiades, at the prime focus of a 15-cm (6-inch) f/4 Newtonian. (b) The North America Nebula through a 135-mm lens. (Jim Baumgardt)*

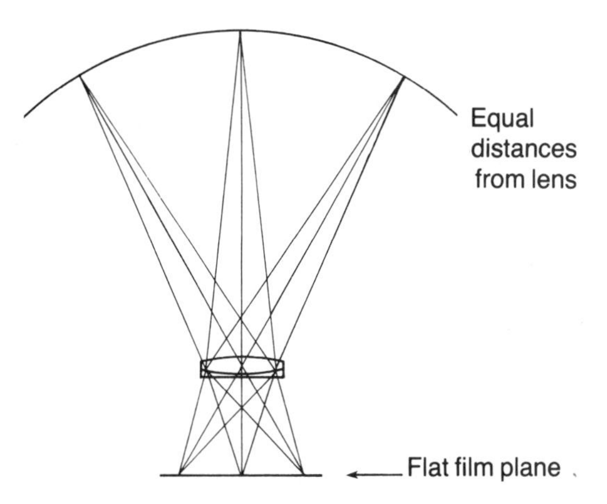

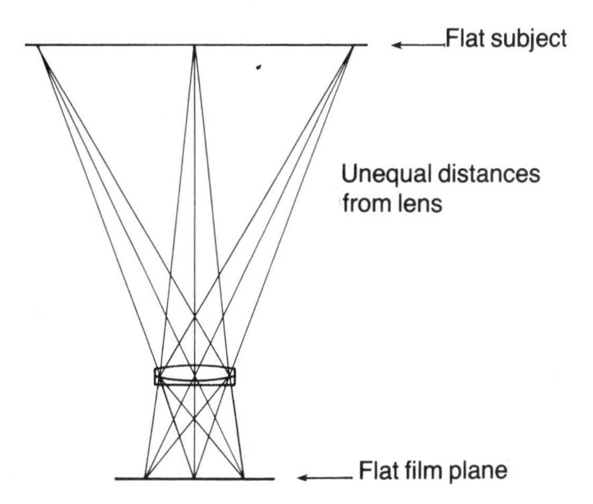

Fig. 8.4. *The field-flattening correction incorporated into enlarging, projection, and copy lenses (bottom) usually makes them unsuitable for astrophotography. A normal lens is shown at the top.*

Fig. 8.5. *An assortment of Zuiko telephoto lenses, and a ×2 teleconverter (shown attached to the camera body). (Olympus Optical Company)*

Fig. 8.6. *Extreme telephoto lenses: 600-mm f/6.5, 1000-mm f/11, 300-mm f/4.5, and 400-mm f/6.3. (Olympus Optical Company)*

priced lenses from other sources. The difference is most marked with normal and wide-angle lenses; it diminishes, often dramatically, with long telephoto and other special-purpose lenses, so that if, for instance, you are looking for a 400-mm telephoto lens, you'll probably have no problems with the least expensive lens available from a reputable dealer. In fact, really bad lenses are rare nowadays; the lowest-priced lenses of today are often adaptations of the 'top of the line' of ten years ago. You can protect yourself to a certain extent by buying from a dealer who will let you return lenses with which you are not pleased after a brief test period.

You can save a good bit of money buying used lenses, but you do take a risk; the previous owner may be selling the lens because its sharpness is not up to standard. The best bargains, and also the lenses least likely to be defective, are those for cameras that are now somewhat out of date; such lenses sell at low prices because of declining demand, and they are available secondhand because their owners traded them in on more modern equipment rather than because of any problems with the lenses themselves. Very specialized lenses (for example, extra-long telephotos) are also good secondhand buys; people buy them for special projects or brief periods of experimentation and decide later to trade them in on something else.

Looseness in the focusing mechanism of a used lens can often be remedied by tightening the tiny screws with a jeweler's screwdriver. A more serious problem is that, over the years, the stop that marks infinity focus may have shifted. If you are fortunate, true infinity focus will still be in the focusing range of the lens, though not at the infinity mark; test and see. If not, you may have to open up the lens mechanism to adjust it.

Any camera lens should of course be equipped with a lens shade and a UV-blocking (Wratten #1A) filter, which, in addition to protecting the front element from dust and scratches, reduces sky fog slightly.

Choosing a telescope

If you are interested in astrophotography but do not yet have a telescope, what kind should you be shopping for? Before I attempt any answers to this question, let me issue two caveats.

One is that amateur astronomers have always placed a high value on making the most of whatever kind of equipment is available. Good astronomical photographs have been obtained with all kinds of telescopes, many of them home-made. To recommend a particular model of telescope as *the* instrument for astrophotography would almost be contrary to the spirit of the hobby.

The other is that, in order to be successful at astrophotography, you need to learn how to use a telescope visually as well. To attempt astrophotography without knowing anything about visual observation would be like trying to photograph birds without taking up birdwatching – you'd never be able to find the objects that you wanted to photograph. So a good case can be made that, especially if you're on a tight budget, your first telescope should be something with which you'll learn about visual observation, even though it may be of little photographic use *per se*. Examples include 5-cm (2-inch) and 6-cm (2.4-inch) refractors and 10-cm (4-inch) or smaller Newtonians, or even a good pair of binoculars. You'll never accomplish much as an astrophotographer unless you become skilled at finding and identifying objects in the sky – a skill you can learn with a small, inexpensive instrument.

Now then. For serious astrophotography, look for a telescope whose manufacturer promotes it as a photographic instrument and has developed a full line of astrophotographic accessories. The 20-cm (8-inch) Schmidt–Cassegrains made by Celestron and Meade are the most popular telescopes in this category; they were designed with photography in mind, and accessories are available to make almost any kind of astrophotography not only possible, but convenient. The Schmidt–Cassegrain optical system gives a sharp image over an exceptionally wide field, and the compactness and portability of the instrument is an asset. Smaller Schmidt–Cassegrains and Maksutov–Cassegrains (down to 9 cm or 3½ inches) offer equal convenience with greater portability and, because they weigh less, do not require as heavy a mount. Naturally, the smaller diameter restricts what you can photograph through the telescope, but it makes no difference for piggy-backing.

A second choice would be a well-made 15- or 20-

cm (6- or 8-inch) Newtonian, preferably one designed for photography (manufacturers include Meade and Edmund). Newtonians are bulkier and less portable than Schmidt–Cassegrains, but they are often much less expensive. Their main disadvantage is that prime focus photography is difficult, since they usually do not provide enough back focus to accommodate a camera body, and in any case the image is not sharp over a large enough area to fill a 35-mm frame.

Refractors can give excellent results, but a refractor large enough for serious work (for example, 10 cm or larger) is usually prohibitively expensive, as well as so massive that it needs to be mounted on a permanent concrete pier or the like.

For photographic purposes, the most important part of a telescope is probably its mount. A shaky telescope is practically useless no matter how good its optics. Moreover, you'll need an equatorial mount with a clock drive for any kind of astrophotography other than the briefest exposures of the sun and moon. A declination slow-motion knob is virtually essential for guiding; a right ascension knob is less important because corrections in right ascension are normally made with an electronic guider.

If you're on a limited budget, a good strategy may be to buy a good tripod and mount, and put a very modest telescope on it (such as a 10-cm or 4-inch Newtonian on a mount designed for a 20-cm (8-in)). You can upgrade to the larger telescope later, and in the meantime get all the benefits of the high-quality mount for piggy-back photography of star fields and the like. In fact, whatever kind of telescope you choose, you may want to order it with a mount normally used on a heavier telescope, or order the lightest of several telescopes that come with the same mount. For example, I chose a 12.5-cm (5-inch) Celestron because it had the same mount as the 20-cm (8-inch) but weighed only half as much and was therefore steadier. (Not that the 20-cm was unacceptable – I just wanted an extra 'edge'. In any case, a sturdier version of the 20-cm has since been introduced.)

Telescopes intended for terrestrial use (for example, target spotting) usually offer little potential for astronomical work; in order to produce a right-side-up image, they use extra lenses or prisms that reduce the quality of the image and the amount of light available. And you'll certainly want to avoid telescopes that are loaded down with unnecessary accessories or whose advertising makes unrealistic claims about magnification ('Amazing 600 power telescope!') – magnifications greater than ×200 are rarely practical with telescopes smaller than 20 cm, and you don't want to deal with manufacturers who are willing to mislead you about what their product will do.

9
Film

How film works

Although other photochemical processes are used in the graphic arts, all present-day photographic films contain crystals of silver halide (silver chloride and/or bromide), which, when treated with suitable developing chemicals, precipitate a dark deposit of metallic silver in proportion to the amount of light to which they have been exposed. The film responds to light even if weeks or months elapse between exposure and development.

Ordinary black-and-white film consists of an emulsion of silver compounds and other sensitizing agents suspended in gelatin and coated on a cellulose acetate (or, sometimes, polyester) base. After exposure, the film is treated first with a developer, which precipitates metallic silver in the exposed areas of the emulsion, and then with a fixer, which washes away the undeveloped silver compounds. The result is a developed negative that is black with silver where light has struck it (corresponding to bright areas in the subject being photographed) and clear elsewhere. When the negative is copied onto another piece of film or paper that works the same way, light and dark are reversed again, and a positive image results.

The new 'chromogenic' black-and-white films (Ilford XP1 and Agfapan Vario-XL) use a refinement of this process: in addition to light-sensitive silver compounds, the emulsion contains dye couplers that, in combination with a special developer, form a tiny colored patch around every grain of precipitated silver. The silver can then be bleached out, leaving only the dye. The obvious advantage of this process is that the silver, a relatively scarce natural resource, can be recovered and reused. The image also looks less grainy because dye spots are less sharp-edged than silver grains, and the film can tolerate larger amounts of overexposure than conventional film, leading to greater exposure latitude.

Dye couplers are the basis of color photography. A color negative film, for instance, has three layers, each sensitive to a particular region of the spectrum (red, green, or blue) and forming an image in a dye of the complementary color (cyan, magenta, or yellow, respectively). The resulting image renders light objects as dark, just like a black-and-white negative, and renders each color as its complement. Copying or printing onto a material that reverses brightness and color a second time results in a realistic positive image. (Color negative films also incorporate a bright orange masking layer, and the response of color enlarging paper is adjusted to compensate for it; hence the colors involved are not exact complements.) The processing of color negative film is hardly any more complex than for black and white; develop, rinse, bleach and fix. In fact, chromogenic black-and-white films can be, and often are, developed in chemicals designed for color negative film.

Things become considerably more complicated if we want slides, that is, film that yields a positive image. The first step is to develop the film normally, so that the silver halide molecules that have been exposed to light are reduced to metallic silver. Then, instead of removing the undeveloped silver halides, we remove the precipitated silver itself. Next, the remaining silver halide crystals are fogged, either chemically or by exposing them to light, and developed; the dark areas on the film at this stage are those on which no light fell originally, so that no silver halide was taken away in the first two steps, and the image is a positive, rendering dark as dark and light as light. After fixing to remove unused silver halides and other chemicals from the emulsion, the image is ready for viewing.

This is known as the black-and-white reversal process, and is used for black-and-white movie film but not, nowadays, for much else. A color version of it is the normal way of processing color slides.

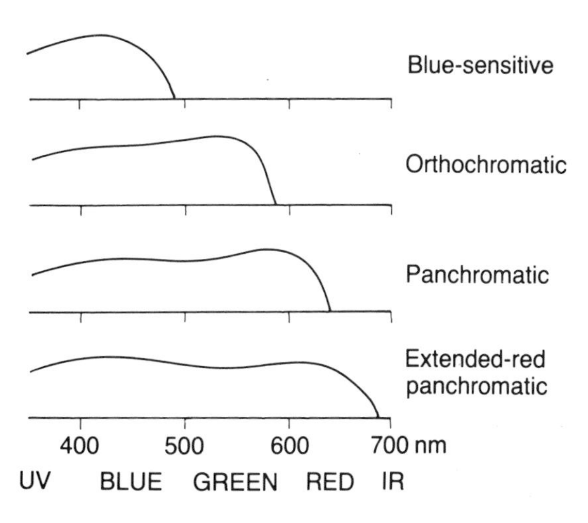

Fig. 9.1. Graphs showing relative spectral sensitivity of blue-sensitive, orthochromatic, panchromatic and extended-red panchromatic film.

Color slide film, like color negative film, consists of three layers sensitive to different parts of the spectrum (red, green, and blue) and forming images in the complementary colors (cyan, magenta, and yellow). Since the final image is a positive, the resulting colors are realistic, in spite of (or rather because of) the fact that complementary dyes are used. For example, photographing a yellow object triggers a response from the red- and green-sensitive layers, causing them to become transparent and leaving the blue-sensitive, yellow-forming layer dark, that is, yellow.

Most color slide films (Agfachrome, Fujichrome, Kodak Ektachrome) use dye couplers to color the images, resulting in relatively simple, non-critical development processes that the hobbyist can easily carry out at home. Kodachrome film, on the other hand, contains no dye couplers, and each of the three layers is processed and dyed separately. Since the processing of the three layers can be adjusted individually, much finer quality control is possible, but because of the equipment required, Kodachrome can be processed only by large laboratories. Kodachrome slides can be expected to last 100 years in storage without fading; Ektachrome and Fujichrome, 50 years.

Spectral sensitivity

Photographic emulsions containing only silver halides are sensitive to wavelengths between about 250 and 500 nanometers, in the ultraviolet, violet, and blue regions of the spectrum (Fig. 9.1). These 'blue-sensitive' or 'unsensitized' emulsions are used on black-and-white photographic paper and certain graphic art films, where their main advantage is that they are unaffected by a red or orange safelight. They have also been used extensively by observatories, mostly in the form of Kodak 103a-O and similar photographic plates; some star catalogues give photographic magnitudes indicating the relative brightness of astronomical objects as photographed on blue-sensitive film. Except for the Kodak 103a-O Spectroscopic film, which is available in long 35-mm rolls for bulk loading, no blue-sensitive materials useful in amateur astrophotography are currently manufactured.

Early in this century, blue-sensitive films and plates were replaced, in most photographic applications, by *orthochromatic* materials – emulsions that respond to all visible wavelengths except red. The results were a gain in speed – if you cover more of the spectrum, you catch more light – and, more importantly, improved photographic realism; there was much less of a tendency for warm-colored objects to photograph too dark. Because the film did not respond to red light, it was still possible to use a safelight when developing it.

By now, though, orthochromatic film has also disappeared, except for certain graphic arts films (Kodalith Ortho, Agfaortho) and scientific emulsions (Kodak's IIIa-J). It has been replaced by *panchromatic* film, which responds to the entire visible spectrum. All general-purpose black-and-white films today are panchromatic; they differ only in their sensitivity to the extreme red portion of the spectrum. Films such as T-Max and Tri-X Pan are sensitized to only part of the visible red range – up to about 640 nm – in order to keep certain red objects, such as people's lips, from looking too light on photographs. Films designed for scientific photography, however, such as Kodak Technical Pan

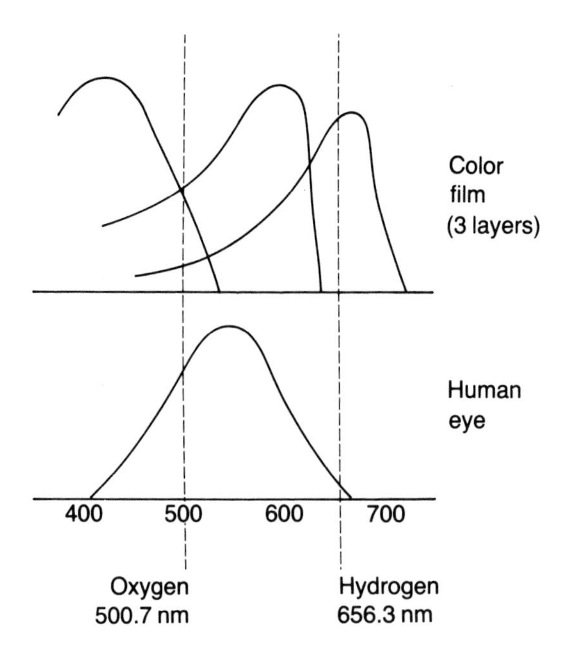

Fig. 9.2. *Why emission nebulae look bluish but photograph as red. Color film responds strongly to the 656.3-nm line of hydrogen (top), but the eye responds almost exclusively to the 500.7-nm line of oxygen (bottom).*

2415, are sensitized to wavelengths as long as 700 nm or a bit beyond. The extended spectral response of these *extended red panchromatic* films increases the effective speed, and also allows scientific photography involving light sources that are almost outside the range of conventional films, such as helium–neon lasers at 633 nm or light-emitting diodes at 650 nm.

The blue-sensitive layer of color film covers about 400 to 525 nanometers; the green-sensitive layer, about 420 to 600; and the red-sensitive layer, about 500 to 690. The wavelength ranges of the three layers overlap somewhat, especially with the faster color films, and the sensitivity of the red-sensitive layer extends farther into the red than that of most black-and-white films. Color films are further classified as daylight type, which reproduces colors more or less as the eye sees them under neutral illumination, and tungsten type, in which the color balance is shifted toward blue to compensate for the relative redness of artificial light. Daylight type film is usually preferable for astrophotography.

Anything that alters the sensitivity of color film – aging, reciprocity failure, or hypersensitization – distorts the color balance insofar as it affects the three layers unequally. Fortunately, astronomers seldom require the degree of realism that is necessary in, say, portrait photography, and most inaccuracies of color can be ignored. The usefulness of color in astrophotography is not that it reproduces the colors seen by the eye – a hopeless endeavor, considering that even visual astronomers cannot agree on them – but rather that, besides enhancing the aesthetic value of the picutre, it makes important detail easier to see.

The most dramatic way in which color astronomical photographs are unrealistic is the way that they render nebulae. Emission nebulae, such as the Orion Nebula, almost always photograph as red, though they look white or pale blue-green to the eye. The reason for this is that such nebulae shine by fluorescence, and their light is confined almost entirely to two narrowly defined wavelengths, 500.7 nm (from ionized oxygen) and 656.3 nm (from hydrogen). The 500.7-nm wavelength is near the maximum sensitivity of the human eye and dominates the visual appearance of the nebula, but it falls between the peaks of the blue- and green-sensitive layers of film, so that color film does not record it well (Fig. 9.2). The 656.3-nm wavelength, on the other hand, records well on the red-sensitive layer, even though the human eye has trouble seeing it. By contrast, reflection nebulae, which shine by reflecting starlight, photograph as white or blue-white. A photograph of an object containing some of each, such as M20 in Sagittarius, can display a dramatic juxtaposition of colors (see for example Plate 7.2).

The characteristic curve

The way a film responds to light is best summarized in its *characteristic curve*, a graph of density versus exposure (Fig. 9.3). The horizontal scale gives the logarithm of the exposure, in meter-candle-seconds; each unit represents ten times as much exposure as the next lower unit, and a one-stop exposure change (multiplying or dividing the exposure by 2) is

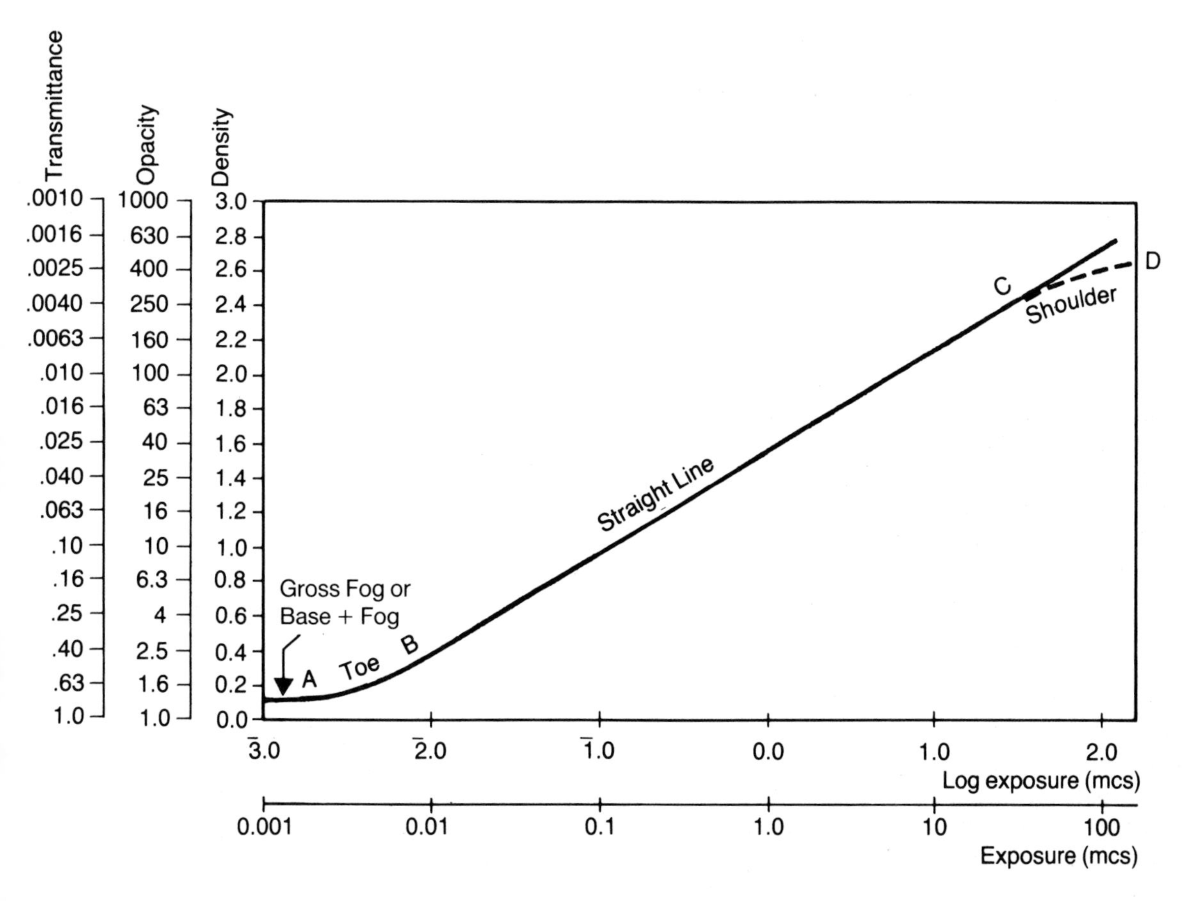

Fig. 9.3. *A film characteristic curve, Exposure, in meter-candle-seconds, is shown on a logarithmic scale. Density is measured logarithmically in the same way as for sun filters. (Eastman Kodak Company)*

equivalent to moving 0.3 of a logarithmic unit to the right or left. The vertical scale is the logarithmic density of the negative, calculated the same way as for sun filters (Chapter 3).

The curve consists of three portions; the toe, where it tapers off to a minimum; the straight-line portion; and the shoulder, where it tapers off to a maximum. In practice, photographs are taken on the toe and the lower part of the straight-line portion; if you go up to the shoulder, black-and-white film becomes quite grainy, but color negative film becomes smoother and shows less grain.

The more steeply the curve slopes, the more contrast the picture will show, because a steeply sloping curve renders a slight exposure difference as a large density difference. The simplest way to measure contrast is to determine the slope of the straight-line portion (that is, how far it moves up as you move one unit to the right). Contrast measured this way is called *gamma* (γ) and is about 0.7 for normal pictorial photography, though much higher values are desirable in astrophotography – on the order of 1.0 to 1.5 for planetary work and as high as 4.0 for star fields. Gamma, of course, varies with development time, but for any given combination of film and developer, there is a maximum, known as *gamma infinity* (γ_∞) beyond which further development does not increase contrast. (Note that the infinity symbol is a subscript; 'gamma infinity' means 'gamma with infinite development' and does not mean that the value of gamma is itself infinite.)

The problem with gamma as a measure of contrast is that it describes only the straight-line portion of the characteristic curve, whereas most photographs use the toe. As an alternative, most Kodak publications use a measure called the *contrast index*, which is the average contrast for the parts of the curve likely to be used in a typical photograph. The ideal contrast index for pictorial photographs is considered to be 0.56.

Films differ as to the shape of the curve; some are 'short-toed' and become quite straight only a short distance above the minimum, while in others, the toe curvature continues a long way up the curve. In pictorial photography, the choice between the two is mostly aesthetic; short-toed films produce a more faithful copy of the tones in the subject, while long-toed films produce pictures in which the contrast varies with the illumination, not unlike the way the human eye sees. In astronomy, the main considerations are that short-toed films are more resistant to sky fog and lens flare, while long-toed films are more likely to pick up faint detail that can be brought out by overdevelopment, intensification, or high-contrast copying.

The shortest-toed general-purpose films available today are the Agfapan Professional series, followed closely by Kodak's Plus-X Pan. Tri-X Pan and Verichrome Pan are medium-toed. The T-Max films and Kodak Technical Pan are medium- to long-toed.

Speed

The speed of a film is its sensitivity to light (and hence the rapidity with which a time exposure can be completed). Under the new International Standards Organization (ISO) system, film speed is reported as two numbers, the first equivalent to the older ASA speed rating, and the second, identified by a degree sign, equivalent to the older DIN rating. Thus Tri-X Pan, formerly ASA 400 and DIN 27, is now ISO 400/27°. The relation between the two numbers is as follows:

$$\text{DIN number} = 1 + 10 \log_{10} (\text{ASA number})$$

ISO speeds indicate the minimum amount of light needed to take a picture that shows adequate shadow detail. They are assigned only to films designed for ordinary pictorial photography, and they presuppose a specific type of processing (for black-and-white films, development to a precisely defined degree of contrast in a special developer similar to D-76). Hence some films do not have ISO ratings, and are assigned an ISO-like speed number, often called Exposure Index (EI), by their manufacturers. Many of the speeds referred to in this book are unofficial estimates rather than actual ISO ratings.

A simple way to measure the approximate speed of a film is to photograph a Kodak gray scale in bright sunlight at *f*/16, and determine the shortest shutter speed at which the two darkest patches are distinguishable on the negative. If the minimum exposure is 1/500 second, then the film is approximately ISO 500; if 1/250, then ISO 250; and so forth.

The true speed of a film varies a bit with the chemicals in which it is developed; as explained in the next chapter, developers that contain Phenidone, such as Ilford Microphen, produce a slight speed increase, and certain fine-grain developers produce a slight speed reduction. Also, the speed of color slide film can be increased by increasing the time in the first developer, thereby developing out more of the silver halide crystals that would otherwise be fogged and darkened by the second developer.

However, the usual practice of 'pushing' black-and-white film by increasing the development time increases the effective speed mostly by raising the

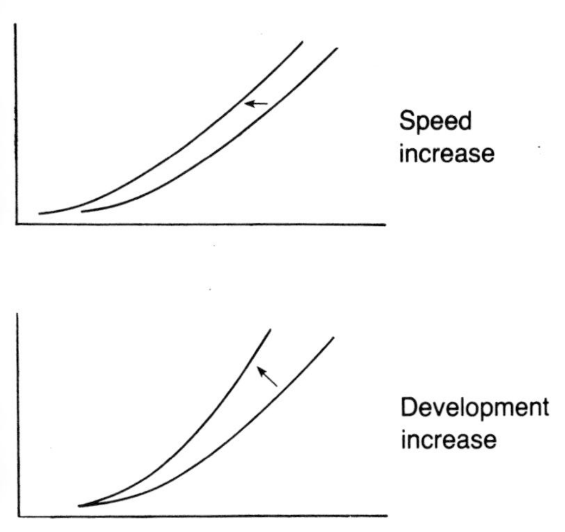

Fig. 9.4. A true speed increase shifts the entire characteristic curve to the left, including the tip of the toe (top). Pushing by increasing development merely makes the curve steeper (bottom). Either way, the film's response to faint light increases.

contrast (Fig. 9.4). Photographs on pushed film are, technically speaking, underexposed, since they lack some shadow detail that would be present on pictures exposed and developed normally. However, the underexposure is compensated by a contrast increase, which increases the density of faint details not actually lost, and the photographer can get reasonably normal-looking negatives while giving the film only half or a quarter of the rated exposure.

Medium- and long-toed films are more pushable than short-toed films, since the effect of increased development on their characteristic curves is more like that of a true speed increase. Kodak T-Max films and Ilford HP5 Plus are often pushed in pictorial photography. Astronomical photographs in which only the faintest detail is of real interest (for example, star fields with nebulosity) should, of course, be pushed generously.

Reciprocity failure

The photochemical law of reciprocity, honored more in the breach than in the observance, states that equal amounts of light energy produce the same effects whether they are delivered slowly or quickly. That is, if a particular light intensity for a particular amount of time yields a properly exposed negative, then twice the light for half the time, or half the light for twice the time, ought to yield the same effect.

Real photographic films do not conform to the law of reciprocity. General-purpose films are most sensitive to light when the exposure time is about 1/100 second, and less sensitive in exposures that are a great deal shorter or longer. The reason for the loss of speed in long time exposures to faint light is that if photons strike a silver atom too infrequently, the effect of each photon may be 'forgotten' before the next one comes along. The short-exposure speed loss (of little concern to astrophotographers) takes place because, if too many photons are delivered to the film at once, not all of them are 'caught'. Reciprocity failure does serve a useful purpose, in that it makes the film insensitive to very weak light leaks, low-level ionizing radiation, and the like; you can safely assume that a light leak that does not fog the film within a few hours will never fog it.

The appropriate form of the classic Schwarzschild formula for calculating the effect of reciprocity failure is:

$$\text{actual speed} = \text{rated speed} \times t^{p-1}$$

where t is the exposure time in seconds and p, known as the Schwarzschild exponent, is a number that varies from film to film – about 0.65 to 0.75 for typical ISO 400 films, 0.8 to 0.9 for films like Technical Pan 2415 and the T-Max series, and 0.95 to 1.0 for a film that shows little reciprocity failure (such as gas-hypersensitized films).

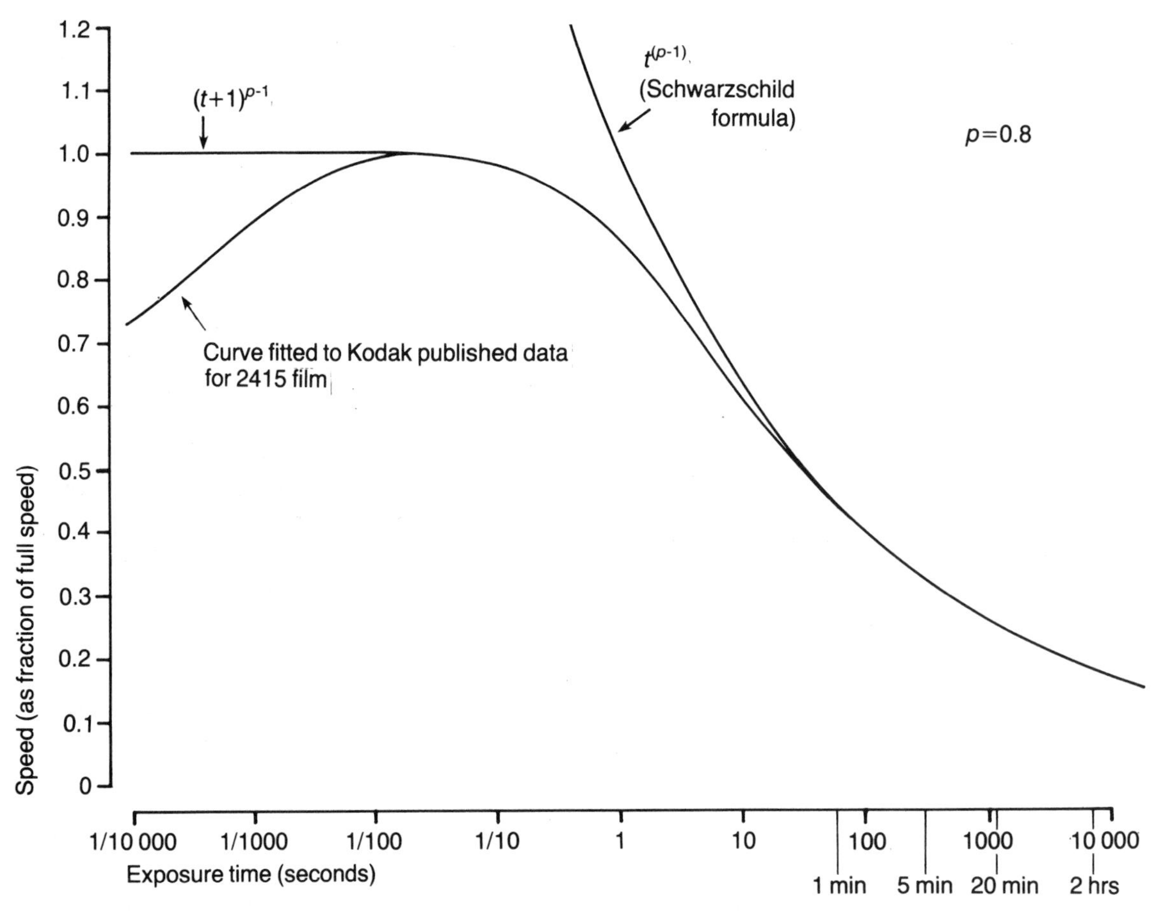

Fig. 9.5. *Reciprocity failure. The modified formula given in the text fits the film's performance more closely than does the Schwarzschild formula.*

As Fig. 9.5 shows, the Schwarzschild formula is reasonably accurate only for relatively long exposures. A much better description of the film's behavior is given by modifying the formula slightly:

$$\text{actual speed} = \text{rated speed} \times (t+1)^{p-1}$$

This makes the calculated curve level off as exposure times approach the optimal range, just as the real curve does. It does not, of course, show the drop-off that occurs with very short exposures.

If, rather than adjusting the film speed, you want to adjust calculated exposure times, use the formula:

$$\text{actual time} = (t+1)^{(1/p)} - 1$$

where t is the exposure time as it would be if the film suffered no reciprocity failure.

Bear in mind that any value of p that you may be working with is quite approximate. The difference between $p = 0.75$ and $p = 0.80$ is equivalent to a 30% difference in film speed in a one-hour exposure – and you can never pin p down even as precisely as that. Calculations of reciprocity failure are hardly more than rough guesses.

With color film, speed loss is not the only problem; there is color shift as well. In general, a shift toward red or yellow can be expected to increase the film's resistance to sky fog, while a shift toward blue can be expected to reduce it, though if the sky fog comes mainly from sodium-vapor streetlights, the opposite may be the case. Table 9.1 summarizes the color shift of various popular films.

Graininess and resolution

When you see grain in a ×10 or ×20 enlargement from a film such as Tri-X, you are not looking at the developed silver grains themselves, but at a mottled pattern resulting from their irregular distribution (measured as *RMS granularity* in manufacturers' data sheets). To see the grains themselves, you would need a microscope. Nonetheless, it remains true that if the grains were smaller or the emulsion thinner, or both, there would be less visible mottle.

High-speed films have coarser grain than slow films, since a faster film requires larger silver halide crystals and a thicker emulsion in order to capture more light. The graininess of any particular kind of film is affected by development; one of the most effective ways to reduce grain is to underdevelop the film slightly. High-energy developers such as D-19 or DK-50 produce grainier pictures than do normal developers such as D-76 (which was classed as a fine-grain developer when it first came out); for finest grain, you can use a developer such as Microdol-X, which results, however, in a slight speed loss.

Although the finest-grained films do have the highest resolving powers, grain is not the only thing that limits film resolution. A chromogenic or color film can have poor resolution even though no grain is visible, because dye clouds do not look as grainy as do silver particles. Even with conventional black-and-white film, a high-contrast picture that shows coarse grain may resolve more detail than a lower-contrast, fine-grained picture.

To see how this is possible, consider what happens as you approach the resolution limit of a film. If a particular film resolves 100 lines per millimeter (lpm), this does not, of course, mean that a 99-lpm test target will be reproduced perfectly while a 101-lpm target will show no detail at all. No; what happens is that as details become finer, they blur together and are reproduced at reduced contrast. The 100-lpm rating means that, at 100 lines per millimeter, the contrast reduction is considered sufficient to greatly impair the usefulness of the image.

This phenomenon is illustrated graphically in Fig. 9.6. What it implies is that increasing the contrast of a picture (by overdevelopment) can increase resolution even if it also increases grain. Also, the same film under the same conditions picks

Table 9.1. *Effect of reciprocity failure on various color films (from the author's tests, 1990)*

Film	Color shifts toward
Ektachrome 100 HC	Little or no change
Ektachrome 200	Yellow (slight)
Kodachrome 200	Red
Kodacolor Gold 1600	Blue, with contrast increase
Fujichrome 100 D	Green (strong)
Fujichrome 400 D	Green-cyan (strong)

Fig. 9.6. *Modulation transfer curve for Tri-X Pan film. The edge effect pushes the response above 100% at frequencies from about 2 to 15 cycles per millimeter – roughly equivalent to the same number of lines per millimeter. (Eastman Kodak Company)*

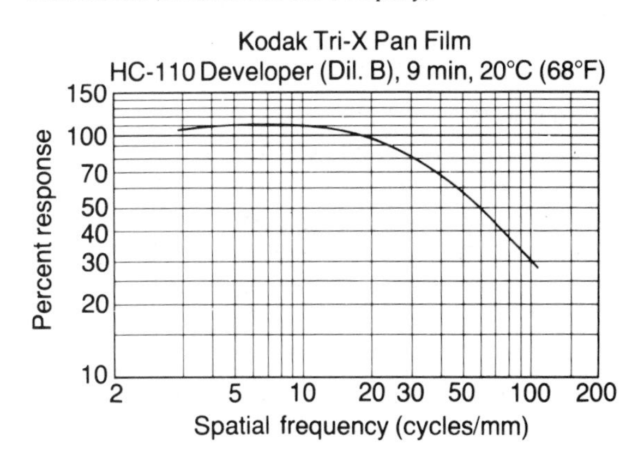

up more detail in a high-contrast subject than in one of lower contrast; hence manufacturers publish both low-contrast and high-contrast resolving-power figures for their films.

Fig. 9.6 shows another peculiarity: the relative contrast actually goes *higher* than 100% at certain spatial frequencies. That is, detail with a certain degree of fineness is actually reproduced at higher contrast than coarser detail.

This results from the *edge effect.* Consider what happens when you develop a piece of film that contains a large light area next to a large dark area. Over the dark area, the developer is consumed rapidly and becomes weak; over the light area, the developer has little work to do and remains strong. In between, near the edge, the developer is stronger than over most of the dark area, but weaker than over most of the light area; hence near the border the light area is lighter, and the dark area is darker, than elsewhere. Agitation reduces the edge effect, as does a strong developer. Lunar photographers sometimes try to enhance the edge effect by using a very dilute developer with little agitation, since increased edge contrast helps bring out lunar detail.

Some specific films

The following recommendations of specific films are based partly on manufacturers' literature, partly on the literature of amateur astronomy, and partly on my own experience. These are not, of course, the only suitable films on the market; film technology is advancing rapidly, and one or two promising new products become available every year. Take these recommendations only as starting points, and do your own comparison tests of products you are curious about. (For extensive test data on Kodak films, see G. T. Keene and M. H. Sewell, 'An evaluation of eight films for astrophotography', *Sky and Telescope,* July 1975, pp. 61–5, and G. P. Brown, G. T. Keene, and A. G. Millikan, 'An evaluation of films for astrophotography', *Sky and Telescope,* May 1980, pp. 433–9. The manufacturers' data sheets for several films are reproduced in Appendix E.)

Kodak Ektachrome 200 is a good general-purpose color-slide film for astrophotography. It suffers moderate reciprocity failure ($p \approx 0.75$) but colors remain fairly accurate in long exposures. If the exposure time is more than a few minutes, Ektachrome 200 is faster than Ektachrome 400 because its reciprocity failure is less severe; it is also finer-grained. The sky background tends to be blue on Ektachrome 200, especially when the film is gas-hypersensitized.

Ektachrome 200 comes in two versions, ordinary and 'Professional'. With the ordinary version, the manufacturer allows for several months of aging between manufacture and development, while the 'Professional' version is designed to be stored under refrigeration and processed promptly. In astrophotography, the ordinary and Professional versions are indistinguishable.

Kodak Ektachrome 100 HC is a fine-grained color slide film with exceptionally vivid colors and, at least in my tests, exceptionally faithful colors in long exposures. It is good for lunar and planetary work and for duplicating slides to increase color saturation and contrast. The reciprocity failure is about the same as Ektachrome 200 ($p \approx 0.75$), and long-exposure star fields with fast lenses have a pleasing appearance with a variety of star colors.

Fujichrome 400 has a varied history. Prior to 1985 it was many astrophotographers' favorite film for star fields, with a somewhat reddish color shift in long exposures (good for eluding sky fog) and good, dense blacks even when hypersensitized. In 1985, it was reformulated and became less suitable for gas hypersensitizing. It may have been modified again around 1989. Today it suffers moderate reciprocity failure ($p \approx 0.75$) with a greenish color shift in long exposures; I have not tested the newest version on the sky. The Fujichrome pictures in this book were taken before 1985.

Like Ektachrome 200, the Fujichromes come in ordinary and 'Professional' versions. Fujichrome films can be processed in Ektachrome E-6 chemistry.

Fujichrome 100 has less reciprocity failure than any other slide film I have tested ($p \approx 0.9$, which is truly impressive). Unfortunately, the colors shift toward green in long exposures, but this is not as catastrophic as it sounds. Star fields are colorful and have a bluish-green background. Fujichrome 100 is a very useful film for deep-sky photography.

Before 1988, Fujichrome 100 star fields had a generally reddish look with vivid star colors, but (probably) somewhat more reciprocity failure. For a comarison of the old and new versions, see the three short articles on color films in *Sky and Telescope*, August 1988, pp. 209–11. The biggest lesson to be learned from Fujichrome is that manufacturers can make significant changes with little fanfare.

Kodacolor Gold 1600 and *Konica SR-V3200* are promising color negative films for deep-sky work, even without hypersensitization. They have less reciprocity failure than most fast films ($p \approx 0.78$ in the case of the Kodacolor) and, although grainy, produce vivid, fairly realistic colors.

Kodak T-Max 100 is outstanding for all types of astrophotography because of its extremely fine grain (comparable to the older Panatomic-X, enlargeable $\times 20$ or more) and very low reciprocity failure ($p \approx 0.88$). Indeed, the T-Max 100 emulsion, on glass plates, is now being used by observatories in place of older Kodak Spectroscopic materials. Its characteristic curve is medium- to long-toed.

For lunar and planetary work, T-Max 100 should be developed according to Kodak's recommendations (see Appendix E), perhaps with push-processing to EI 400 to increase contrast. For maximum speed and contrast in deep-sky work, it can be developed for 10 minutes in HC-110 (dilution A) at 20 °C (68 °F). This increases grain but brings out faint detail.

The main drawback of T-Max 100 is that, like nearly all black-and-white films, it has virtually no response to the 656-nm emission from ionized hydrogen, and hence does not photograph emission nebulae very well.

Kodak T-Max 400 and *T-Max P3200* are faster black-and-white films with good reciprocity characteristics ($p \approx 0.80$ in both cases). T-Max 400 is somewhat finer-grained than Tri-X Pan; T-Max P3200 is grainier than Tri-X, but nowhere near as grainy as the old Kodak 2475 Recording Film. Their characteristic curves are long-toed and both films are quite pushable. T-Max 400 and P3200 are useful for long exposures of deep-sky objects when grain is tolerable and hypersensitized film is not available.

Kodak Tri-X Pan and *Ilford HP5 Plus* represent the older generation of ISO 400 black-and-white films, with medium-toed characteristic curves (quite unlike T-Max 400) and fairly severe reciprocity failure ($p \approx 0.65$ to 0.70). Precisely because of the reciprocity failure, these films are good for meteor photography. Also, they are relatively tolerant of incorrect exposure and incorrect development time.

Agfapan 400 is quite different from its Kodak and Ilford competitors. Its very-short-toed characteristic curve makes it seem slower than Tri-X Pan or T-Max 400. It is also finer-grained and requires longer development times in the same developers. Reciprocity failure is fairly severe ($p \approx 0.65$ to 0.7).

Ilford XP1 is a chromogenic black-and-white film and is developed in the same chemicals as Kodacolor. Because the image is composed of dye rather than silver granules, pictures are surprisingly free of grain, even though the resolving power is no greater than with other fast films.

XP1 has great exposure latitude and is ideal for high-magnification lunar work and lunar and solar eclipses. The characteristic curve has an unusual shape with more contrast in the shadows than in the highlights. The low contrast makes XP1 less than ideal for planetary photography. In deep-sky work, reciprocity failure is fairly severe ($p \approx 0.7$), but this is made up for by the fact that the true speed of the film is very high (perhaps ISO 1600). The absence of distracting grain is a benefit. See John R. Sanford, 'Testing a chromogenic film for astrophotography', *Journal of the British Astronomical Association* 92.4 (1982) pp. 196–7.

Kodak Technical Pan Film (2415, 4415, or 6415 depending on size) is Kodak's all-purpose, extremely fine-grained black-and-white film for scientific photography. Originally designated 'SO-115', it was introduced to replace, among other things, Kodak High Contrast Copy Film, which in turn replaced an even older material called Microfile.

Technical Pan is so fine-grained that grain is rarely visible in a picture; ×25 enlargements are quite feasible if the image is sharp enough. Processed in different developers, Technical Pan can take the place of several different films; it acts as an ultra-fine-grain pictorial film when developed in Technidol, a high-contrast scientific film when developed in HC-110, or a graphic-arts-type copying medium, with very high contrast, when developed in a print developer such as Dektol. The effective speed varies from 25 to 200 depending on development.

Technical Pan 2415 is supplied on a polyester (ESTAR) base, rather than the usual cellulose acetate, and is not easy to tear. If you are accustomed to tearing the film loose after pulling it out of the cartridge for development, bring along a pair of scissors; trying to rip the film with your fingers will do no more than stretch it slightly. Also, the film base is thinner than that of most films and may require padding in some cameras.

Gas-hypersensitized Technical Pan is exceptionally easy to use; it retains its speed and remains relatively free from fog even if kept out of refrigeration for several days. Even without hypersensitization, Technical Pan can yield pleasing results in deep-sky exposures of 10 minutes or more; the striking resolution and freedom from grain give the picture a large-format look that more than makes up for the lack of faint detail.

With its extended red sensitivity, Technical Pan responds very well to the 656-nm emission from hydrogen nebulae. This is one reason that deep-sky photographs taken on this film look so dramatic. Further, the exposures can be made through deep red filters (Wratten #25, #29) to eliminate sky fog even in urban settings.

Like T-Max 100, Technical Pan is now supplied to observatories on glass plates.

Kodak Spectroscopic Films are special black-and-white products manufactured for scientific research. The ones of interest to astrophotographers are those designated with the suffix 'a' (103a-O, IIa-O, IIIa-J, and the like, as opposed to 103-O or III-J), since they are specially treated to reduce reciprocity failure.

In their time, the Spectroscopic films and plates were a great advance and resulted in many scientific discoveries. With the advent of gas-hypersensitized Technical Pan, however, they are almost obsolete. In fact, even without hypersensitization, Technical Pan and T-Max 100 outperform all the Spectroscopic emulsions in speed, grain, or both.

Kodak has always supplied Spectroscopic films on special order only, with large minimum orders, and the films available at any given time have varied. Here are details on several of the most widely used.

Kodak Spectroscopic Film IIIa-J is a fine-grained film with roughly orthochromatic sensitivity; without gas hypersensitization, it is comparable to Technical Pan, and gas hypersensitization can increase its speed by a factor of 6 or more. Type 103a film, which is coarse-grained by fairly fast, has been available in several different spectral sensitizings: 103a-O is blue-sensitive, 103a-G is approximately orthochromatic, 103a-E is panchromatic with a peak at the red end of the spectrum, and 103a-F is more like conventional panchromatic film. IIa-O, which is blue-sensitive, is slightly slower than 103a-O but almost as fine-grained as IIIa-J.

Bulk loading

A good way to save money on any kind of 35-mm film is to purchase it in 50- or 100-foot (15- or 30.5 meter) rolls and load it into individual cartridges yourself. With the aid of a gadget called a *bulk loader,* you can do this in full room light; total darkness is required only when you place the long roll into the bulk loader at the start.

The cheapest kind of bulk loader passes the film through a plush felt-lined light trap. Unfortunately, a grain of sand or similar particle trapped in the felt can scratch 100 feet of film before you realize that anything is wrong. The alternatives are to use a more expensive bulk loader with a light trap that opens up and allows the film to pass freely when the outer enclosure is closed, or even to treat the felt-type bulk loader (under $10) as a disposable item, to be replaced or retired to a less critical application after only a couple of rolls.

The other problem you have to face is that of light leaks, which are usually the fault of the cartridge. I have had excellent results with Kodak Snap-Cap Film Cartridges, which are fairly expensive, and with reused Ilford and Agfa film cartridges, but not with the cheap off-brand cartridges that are sometimes provided free with the bulk loader. It is important not to use a cartridge more than two or three times, and never to use a catridge that is in any way suspect.

A roll of 35-mm film requires 38 mm (1.5 inches) of film for each frame, plus about 230 mm (9 inches or six frames) for leader and trailer. A 100-foot (30.5-meter) roll makes nineteen 36-exposure rolls, thirty 20-exposure rolls, or forty-four 12-exposure rolls.

Standard bulk loaders accommodate 100-foot rolls of acetate-base film or 150-foot (45-meter) rolls of polyester-base film, which are the same diameter. These long rolls are designed to fit 35-mm movie cameras and industrial cameras that must take thousands of pictures on a roll, so they exist in many varieties. Fortunately, you seldom need to do more than ask for film in a format suitable for bulk loading. If you do have to specify a format exactly, however, these are the things you should know:

1 The film must be perforated; few cameras take unperforated film. There are two standard kinds of perforations, KS (Kodak Standard), which are rectangles with rounded corners, and BH (Bell and Howell), which are rectangles with the two short sides bulging. Cameras will, in general, take either.

2 The film is normally wound on a core; any kind of core that has a central hole at least 6 mm (1/4 inch) across will do. Film wound on a spool (which, unlike a core, has sides) may not fit into the bulk loader.

10
Developing, printing, and copying

The darkroom

There are many advantages to doing your own darkroom work. It gives you complete control over the appearance of your pictures (control that you won't want to relinquish, once you've experienced it), it gives you opportunities to experiment, and, with black and white at least, it costs so much less than commercial processing that the money saved over ten or fifteen rolls of film can pay for the darkroom equipment. (If you don't want even to consider doing your own darkroom work, skip ahead to the section on slide duplication later in this chapter; it tells you how to get many of the same benefits while relying on commercial photofinishers.)

There are many good books on how to outfit and use a darkroom, so I won't try to cover the whole subject here. Instead, let me outline what darkroom work is like for the benefit of readers who are unacquainted with it, and then give some specific practical hints.

A darkroom for developing 35-mm film is actually dark only for the few minutes that it takes to load the film into a light-tight developing tank (a procedure that can, of course, be carried out anywhere that complete darkness can be obtained). Once the film is in the tank, the lights are turned on; chemicals can be poured in and out of the tank in full room light. The chemicals are a developer, whose temperature is brought to 20 °C (68 °F) by placing the container in a basin of warm or cool water; a water rinse or stop bath; and a fixer. The whole process takes perhaps twenty or thirty minutes, after which the tank is opened and the film is washed for perhaps half an hour, then hung up to dry.

Printing is done under an orange safelight that does not affect the photographic paper. Although a negative can be placed directly against the paper and *contact printed* to produce a print the same size, most people use an *enlarger*, which projects an image of the negative (in white light) onto a larger piece of paper. The exposure is controlled by adjusting the enlarger's *f*-stop and varying the exposure time, which is of the order of 10 seconds; correct exposure is usually determined by trial and error on small scraps of paper. The exposed paper is processed in trays of developer, stop bath, and fixer, in succession; modern resin-coated paper then requires only about five minutes of washing before being hung up to dry.

That is the basic outline; for more information, see *Basic Developing, Printing, Enlarging in Black and White* (Eastman Kodak Company, 1982) or any of a variety of other books available at camera shops. Now for some practical hints based on my own experience.

If you have trouble finding a suitable location for a darkroom, remember the option of working only at night – not having daylight to deal with makes light-proofing much easier. You can also do without running water, since you can wash film and prints elsewhere.

Test the light-tightness of a darkroom for film developing by going into it and letting your eyes adapt to the darkness. If, after ten minutes, you still cannot see your hand in front of you, all is well; any faint glimmers that may remain will probably have no photographic effect. (But a digital wristwatch that lights up when a button is pushed can be a hidden peril.) Somewhat more stray light is tolerable in printing than in film developing, especially if each individual piece of paper is not out in the open for very long.

I greatly prefer chemicals in liquid concentrate form rather than powders. Powdered chemicals are difficult to mix and require large amounts of storage space once prepared; also, chemical dust, especially fixer, can easily damage photographic materials.

With liquid concentrates, on the other hand, you can mix up only as much as is needed and throw it away after one use so that you never use anything that isn't fresh; moreover, you don't have to filter the solution to remove undissolved powder. I use Kodak T-Max and HC-110 film developers (see next section), Ilford Universal print developer, diluted acetic acid for a stop bath, Kodafix solution (a rapid-acting fixer) for film, and Ilford Universal fixer for prints. If you do a lot of darkroom work (say, at least one session per week throughout the year), you may want to consider the Kodak Ektaflo line of liquid concentrate chemicals (print developer, stop bath, and fixer), which is supplied in one-gallon collapsible plastic containers.

Water that is fit for human consumption and free of visible particles is usually good enough for photographic chemicals; variations in the acidity of the water can, however, alter the strength of a developer. If your faucet has an aerator, remove it before mixing developers that are to be stored for several days or more; air dissolved in the water reduces the developer's storage life. Prepared developer can be protected from air with a chemically inactive gas (such as 'Dust-Off', available in spray cans) or by blowing some carbon dioxide (from your lungs) through a tube into the airspace at the top of the bottle. (This latter technique is also used in France to preserve wine.)

Processing black-and-white film

Most of the characteristics of a film, including its speed and graininess and the shape of its characteristic curve, are affected by the way it is developed. The important variables are the choice of developer and the amount of development time.

Ordinary black-and-white film developers contain one or more developing agents, plus an alkali such as sodium carbonate (the 'accelerator'), a preservative such as sodium sulfite, and a restrainer such as potassium bromide that keeps the developer from fogging the unexposed areas of the film.

The characteristics of a developer depend largely on the developing agent or agents that it uses. The three most popular are hydroquinone, Metol (para-methylaminophenol, also known as Elon), and Phenidone (1-phenyl-3-pyrazolidinone).

By itself, hydroquinone produces high contrast and high film speed; it is the sole developing agent in graphic arts developers such as Kodak's D-8. Pure hydroquinone developers result in contrast too high for most photographic applications, however, and hydroquinone is normally used in combination with another developing agent. An example is D-19, Kodak's standard developer for scientific photography, which uses hydroquinone with some Metol added; it gives relatively high contrast, high speed, and good freedom from fog.

Adding a still higher proportion of Metol results in a good general-purpose developer such as Kodak's D-76 and Ilford's ID-11. With all Metol and no hydroquinone, you get a fine-grain developer such as Kodak's D-23 and Microdol-X or Ilford's Perceptol; contrast and film speed are slightly lower than normal, graininess is markedly reduced, and exposure latitude is increased.

Phenidone gives high speed and low contrast, an unusual combination. By itself, in a developer such as Kodak Technidol LC powder (now discontinued), Phenidone produces normal pictorial contrast on intrinsically high-contrast films such as Technical Pan. (Kodak Technidol Liquid gives similar results but contains hydroquinone as well as Phenidone.) In combination with hydroquinone, Phenidone produces normal contrast with maximum film speed. The classic developer of this type is Ilford Microphen, which is good for push-processing but tends to produce excessive fog on gas-hypersensitized film.

My favorite speed-increasing developer is Kodak T-Max developer, which appears to be a Phenidone–

Dilution	Kodak's method		Author's method	
	Dilute stock solution	Amount of stock solution to make 240 ml	Dilute concen- trate	Amount of con- centrate to make 240 ml
A	1:3	60 ml	1:15	15 ml
B	1:7	30 ml	1:31	7.5 ml
D	1:9	24 ml	1:39	6.0 ml
F	1:19	12 ml	1:79	3.0 ml

Table 10.1. *HC-110 dilutions*

The stock solution is prepared by diluting the concentrate 1:3 with water; 240 ml is the amount of developer needed to fill a standard 35-mm film developing tank.

hydroquinone combination although its formula has not been made public. T-Max developer is a liquid concentrate that is normally diluted 1:5 but can be diluted 1:7 or 1:9 for economy and to enhance edge effects. The concentrate can be stored for months and diluted as needed. Kodak T-Max RS developer, a very similar product, is less convenient because the concentrate comes in two parts which must be mixed and diluted all at once. See Appendix E for development times and further information.

Perhaps the most versatile developer is Kodak HC-110, whose formula has not been made public; it gives results similar to D-76 or, in higher concentrations, D-19, with somewhat finer grain. HC-110 is supplied as a syrupy liquid, and Kodak recommends that you dilute the whole bottle of concentrate 1:3 to make a stock solution, from which you then make up the working solution as needed. The rationale is that the liquid is too syrupy to measure accurately. However, I have no trouble measuring it with a syringe designed for giving liquid medicines to babies; an accuracy of ±0.2 ml is sufficient. The 16-ounce bottle of concentrate is enough to develop dozens of rolls of film.

HC-110 works well with gas-hypersensitized film and, indeed, with all films for which D-19 is recommended. Development for 10 minutes in HC-110 (dilution A) is about equivalent to 4 or 5 minutes in D-19. There is no developer fog even with prolonged development.

Table 10.1 gives instructions for preparing the most commonly used dilutions of HC-110, both Kodak's way and my way, and Table 10.2 lists

Film	Effective speed (ISO)	Development time (minutes) HC-110 (B) 20 °C (68 °F)	T-Max (1:4) 24 °C (75 °F)
Technical Pan	100	6	–
T-Max 100	100	7	6½
T-Max 100	400	9½	9
T-Max 400	400	6	8
T-Max 400	1600	8	8½
T-Max P3200	800	8	6½
T-Max P3200	3200	11½	9½
Plus-X Pan	125	5	5
Tri-X Pan	400	6½	5½
Ilford HP5 Plus	400	5	5
Fuji Neopan 400	400	4	4½
Fuji Neopan 1600	1600	7	4
Agfapan 400	400	8	–

Table 10.2. *Development times for various films in Kodak HC-110 and T-Max developers*

Table 10.3. *Equivalent development times, in minutes, at different temperatures*

Find the specified temperature at the top; find the specified time directly below it; and then use other values in the same row to determine development time at slightly different temperatures.

19 °C (66 °F)	20 °C (68 °F)	21 °C (70 °F)	22 °C (72 °F)	23 °C (74 °F)	
5.5	5	4.6	4.2	3.8	minutes
6.6	6	5.5	5	4.6	minutes
7.7	7	6.4	5.8	5.3	minutes
8.7	8	7.3	6.7	6.1	minutes
9.8	9	8.2	7.5	6.9	minutes
11	10	9.1	8.4	7.6	minutes
13	12	11	10	9.1	minutes
15	14	13	12	11	minutes
18	16	15	13	12	minutes
20	18	16	15	14	minutes

To perform other conversions, multiply the specified development time by $\exp(-0.081\ T)$, where T is the temperature difference in degrees Centigrade (positive if higher than specified temperature, negative if lower). For Fahrenheit multiply by $\exp(-0.045\ T)$.

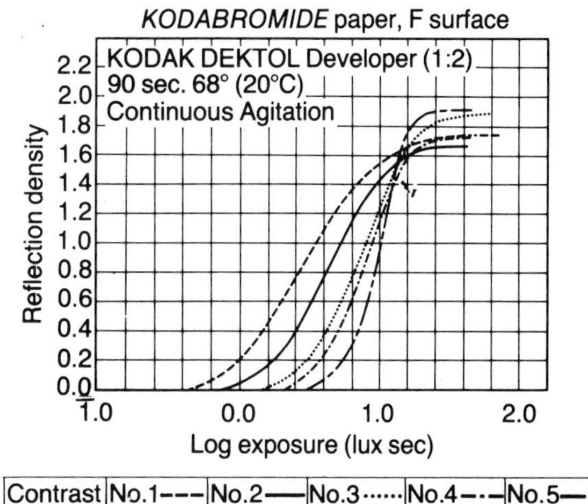

Fig. 10.1. Characteristic curves of the five contrast grades of Kodak Kodabromide enlarging paper. (Eastman Kodak Company)

development times for a number of films, from various sources. See also Appendix E.

A few other developing agents are in use. An example is Agfa Rodinal, which goes back to the early days of photography; it uses para-aminophenol in a strongly alkaline solution and is supplied as a liquid concentrate that can be diluted as much as 1:100 to yield normal contrast, normal speed and enhanced edge effects.

Besides increasing the edge effect, highly diluted developers change the shape of the film's characteristic curve, reducing contrast in strongly exposed areas. This is known as *compensating action*; like the edge effect, it results from exhaustion of the developer over large dense regions. Compensating development is helpful in photographs whose highlights tend to be too light and void of detail, such as pictures of the moon (especially near quarter phase) and of nebulae with overexposed central regions. It is important not to agitate too much, and to use a large volume of developer (for example, developing one reel in a full two-reel tank) so that an adequate amount of developing agent is available. For fuller information, see Ansel Adams, *The Negative* (Boston: Little, Brown and Company, 1981), pp. 226–32.

Because they are printed on high-contrast paper, astronimcal negatives are especially sensitive to streaking that results from improper agitation. In general, too little agitation causes streaks that seem to flow from bright or dark objects in the picture, while streaks coming from 35-mm film perforations result from excessively monotonous agitation. Fogging around the sprocket holes can also result from light leaks or from rewinding the film too tightly in the cartridge.

I get more uniform development with Paterson plastic tanks than with metal tanks. During development, I agitate the tank every 30 seconds by inverting it quickly and vigorously several times.

The amount of development that takes place depends, of course, on both the duration and the temperature of development. To some extent, changes in temperature can be made up for by changes in time; Table 10.3 is a brief conversion chart for most commonly used developers. The conversions are not exact, however, and it is better to standardize developing temperature.

After development, the film goes into a stop bath, which can be plain water or a solution of not more than 2% acetic acid, and then into a fixer. The fixing time depends on the type of fixer and the thickness of the film's emulsion; consult the instructions for each type of film. Washing takes about 30 minutes.

Black-and-white printing

Enlarging paper comes in contrast grades numbered from 0 or 1 to 5 (see Fig. 10.1), with grade 2 or 3 considered 'normal'. Astronomical photographs, particularly of planets and star fields, benefit from printing on grade 4 or 5; Kodak Kodabrome II RC paper, Grade 5, and Agfa Brovira and Brovira-Speed Grade 5 (BEH), give especially high contrast. High-contrast paper is relatively intolerant of exposure errors during enlarging, and tends to bring out film grain. Still, you get less grain by printing a normally developed negative on high-contrast paper than if you tried to achieve the same contrast increase by overdevelopment.

On the other hand, you do occasionally need paper in the lower contrast grades. One possible approach is to use variable-contrast paper (Kodak

Polycontrast and Polyprint or Ilford Multigrade), which gives you contrast grades 0 to 5, in half-grade steps, when exposed through different filters. The paper has two emulsions, a high-contrast emulsion sensitive to blue light, and a low-contrast emulsion sensitive to green light. You place a yellow (blue-blocking) filter in the enlarger to use the low-contrast emulsion, a magenta (green-blocking) filter for the high-contrast emulsion, or various immediate filters to use both emulsions in combination. Of the variable-contrast papers, Kodak Polycontrast III RC gives the greatest contrast range.

With some papers, such as Kodak Polyprint, it is easy to adjust the density of the print by varying the developing time; you simply leave the print in the developer until it looks right, then transfer it to the stop bath. Unfortunately, prints that 'look right' in the dim environment of the darkroom are usually too light; you find yourself wanting to remove the print from the developer too soon, resulting in weak, mottled blacks. Unlike Polyprint, most other resin-coated papers have the developing agent in the emulsion to facilitate machine processing. With such papers you cannot vary the degree of development.

A good print uses the entire density range of the paper, from white to black. As Fig. 10.1 shows, paper is contrastier than film, and, since its entire characteristic curve is being used, there is no room for exposure error. In fact, pictorial photographers routinely 'dodge' or 'burn in' areas of the print, controlling their exposure selectively to get them within range. I find that lunar photographs (other than full moon) and pictures of nebulae with overexposed central regions benefit from burning-in; most other astronomical photographs do not.

The main difference between paper developers and film developers is that paper developers contain large amounts of alkaline accelerator to make them work quickly. If you develop film in a paper developer, you get unacceptably coarse grain – unless, of course, the film is an extremely fine-grained one such as Kodak Technical Pan, in which case you get very quick development and high contrast.

As a rule, print developers are much more interchangeable than film developers; you can develop practically any paper in practically any print developer without noticing more than a slight change in contrast or speed.

Prints of star fields need to be perfectly smooth and free of mottle; making them so requires more than the usual amount of care. Paper and chemicals must be fresh. Each print must be developed at least 1 minute with good agitation and then given at least 20 seconds in an acid stop bath (1.5% acetic acid, or, to avoid the odor, 1% citric acid). Without the acid stop bath, the action of the fixer would be impaired.

White lights must not be turned on until the print is fully fixed. The is not as bad as it sounds; Ilford Universal Fixer will fix all resin-coated papers in less than 1 minute. So will Kodak Rapid Fixer if the hardener (Part B) is left out. Film benefits from having its emulsion hardened, but paper does not; a non-hardening fixer works faster and washes out more easily.

Combination printing of several exposures of a planet was dealt with in Chapter 6. Another powerful technique for improving picture quality is *unsharp masking*, a process that amounts to 'automatic dodging' since it lightens and darkens parts of the picture to keep them within the response range of the paper.

To understand unsharp masking, imagine what would happen if you took a negative, made a low-contrast contact print of it on film, and sandwiched the two. You'd get a contrast reduction, of course; the contact print would partly cancel out the original negative. This is a practical technique and is

a

b

Fig. 10.2. *A 25-minute exposure of the Orion Nebula on 103a-F film at the prime focus of a 20-cm (8-inch) f/5 Newtonian. (a) High-contrast print of the original negative; (b) the result of unsharp masking. (Dale Lightfoot)*

used in some color copying processes.

If the contact print is blurred so that it does not pick up fine detail, an interesting thing happens. The densities of large areas are brought towards the average, but the contrast of the fine detail is not reduced. The effect is the same as if the print had been made by someone with great, indeed superhuman, dodging skill, and the detail that shows up in nebulae and similar objects is remarkable (Fig. 10.2).

The best way to make an unsharp mask is to interpose a thin piece of glass, not more than a couple of millimeters thick, between the negative and the film, then use a light source (such as an enlarger) a meter or so away to make the exposure. You can use conventional black-and-white film if you are willing to work in total darkness; an alternative is to use Kodak Fine Grain Positive Film (in 8×10 sheets) and Eastman Fine Grain Release Positive Film 5302 (in long 35-mm rolls), which consist of an enlarging-paper-type emulsion on a film base. (If you try to make an unsharp mask by putting an enlarger or slide duplicator out of focus, you'll have difficulty getting the images exactly the same size; the image size shifts with a change of focus.) For more on unsharp masking, see Dale Lightfoot, 'Making the most of black-and-white astronegatives', *Astronomy,* January 1982, pp. 51–5.

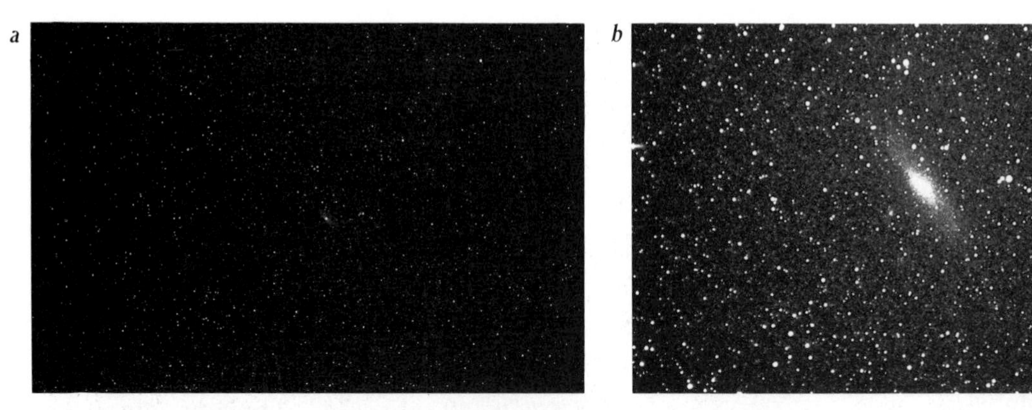

Fig.10.3. *The galaxy M31 in Andromeda, 20 minutes on Ektachrome 200 with a 100-mm f/2.8 lens. (a) Original slide; (b) enlarged duplicate on Kodachrome 25, exposed two stops more than the meter indicated. (By the author)*

Slide duplication

After the camera and the telescope, the third most important basic tool for astrophotography may well be a slide duplicator. With it you can alter the contrast, magnification, and color balance of images, often bringing out hidden detail dramatically (Plate 10.1), without having to do any developing or printing yourself.

There are four main types of slide duplicator:

1 A cylindrical attachment about 50 mm (2 inches) long, containing a positive lens of about 50 mm focal length, that fits onto the front of your 50-mm lens.

2 A long cylindrical device that contains its own copying lens (usually very slow, for example *f*/22) and fits onto a camera body with a T-adapter.

3 A camera, with bellows or extension tubes and suitable lens, taking a picture of a slide on a light box in a darkened room.

4 A light-tight slide holder that fits on the front of a bellows, around and in front of the lens, and uses the same optical system as 3 but is immune to glare.

Setups 1 and 2 are typically quite inexpensive; I much prefer 1, since it gives you a brighter image to focus on. The advantage of 2 is that it can give you variable magnification (but then, so can 1 if you put it on a telephoto or zoom lens). For a long time, I used a slide duplicator of the type 2 that was not supposed to require focusing; after getting some quite out-of-focus results, I made a set of plastic shims with which to move the slide forward or backward in the mount.

Duplicators of type 3 or 4 may easily cost more than the camera; 3 can, however, be affordable if you build the light box and stand yourself. Commercial light boxes often include carefully controlled light sources with room for filters.

Paradoxically, it is much easier to improve a slide by duplication than to make a copy that looks just like the original. If you want copies of existing slides, you're much better off sending them to Kodak or Kodalux. The usefulness of a slide duplicator lies in the changes that the image undergoes when it is duplicated.

When you copy a slide onto ordinary color slide film, it undergoes a contrast increase. If you want an exact copy of the original, you must prevent the contrast increase, either by prefogging the film slightly or by using a special slide duplicating film. With astronomical photographs, however, the contrast increase is almost always beneficial; it helps to bring out faint stars and make planetary detail more visible. The film onto which you duplicate should, of course, be fine-grained; I use either Kodachrome 25 or, for greater color separation, Ektachrome 100 HC.

For determining exposures, through-the-lens metering in the camera is practically essential; automatic exposure is helpful provided some amount of override is available. Usually, the exposure indicated by the meter results in a duplicate that is in about the same density range as the original, although, because of the contrast increase, the darkest parts of the picture are black and the lightest parts are washed out. (Surprisingly, this holds true even with pictures of celestial objects

Fig. 10.4. The galaxies M81 and M82: 28 minutes on 103a-F film at the prime focus of a 20-cm (8-inch) f/5 Newtonian. (a) Normal print on high-contrast paper; (b) result of rephotographing the original print; (c) result of contact-printing the original negative on very-high-contrast graphic arts film, then contact-printing the resulting positive onto the same material to produce the negative from which the final print was made. (Dale Lightfoot)

on a black background, where you'd expect an averaging meter to give incorrect results.) With a planetary photograph, where all the detail is in the middle of the density range, this may be just what you need.

Much of the detail of interest in a star-field photograph is present on the slide but too dense to see unless you view the slide by a very strong light. It is in such cases that the slide duplicator proves its worth: you can lighten such a slide by exposing about two stops more than the meter indicates. In either case, bracket exposures widely.

Naturally, the color balance of a slide duplicate depends on the light source used. I normally use blue photofloods for duplicating onto daylight film; electronic flash or sunlight also works well. Ordinary tungsten light (from a slide projector, enlarger, or ordinary light bulb) produces a color shift toward coppery-red that is sometimes helpful in eliminating blue or blue-green fog and bringing out reddish nebulosity.

A slide duplicator can also substitute for an enlarger; you can put a negative in a slide mount and use the slide duplicator to make positives from it. You need, of course, a film that has the contrast of enlarging paper. In black and white, Technical Pan 2415, exposed at ISO 200 and developed 12 minutes in HC-110 (dilution B), or developed to saturation (about 5 minutes) in a paper developer, should make a good starting point for experimentation. To make positives from color negatives, you can use Kodak Vericolor Slide Film 5072, which is designed especially for this purpose.

Rephotography

Rephotography is like slide duplication except that you copy, not a slide, but a print – usually a carefully made glossy 8×10. A copy stand is helpful but not essential; a table with a couple of flexible-arm lamps for shadow-free illumination will suffice. To bring out faint detail on a picture using this technique, make a slightly lighter-than-normal print (you don't want anything to disappear into the shadows) and photograph it on Technical Pan 2415 or a similar high-contrast film. Fig. 10.4 shows what rephotographing can accomplish.

Rephotography is the key to Jim Baumgardt's

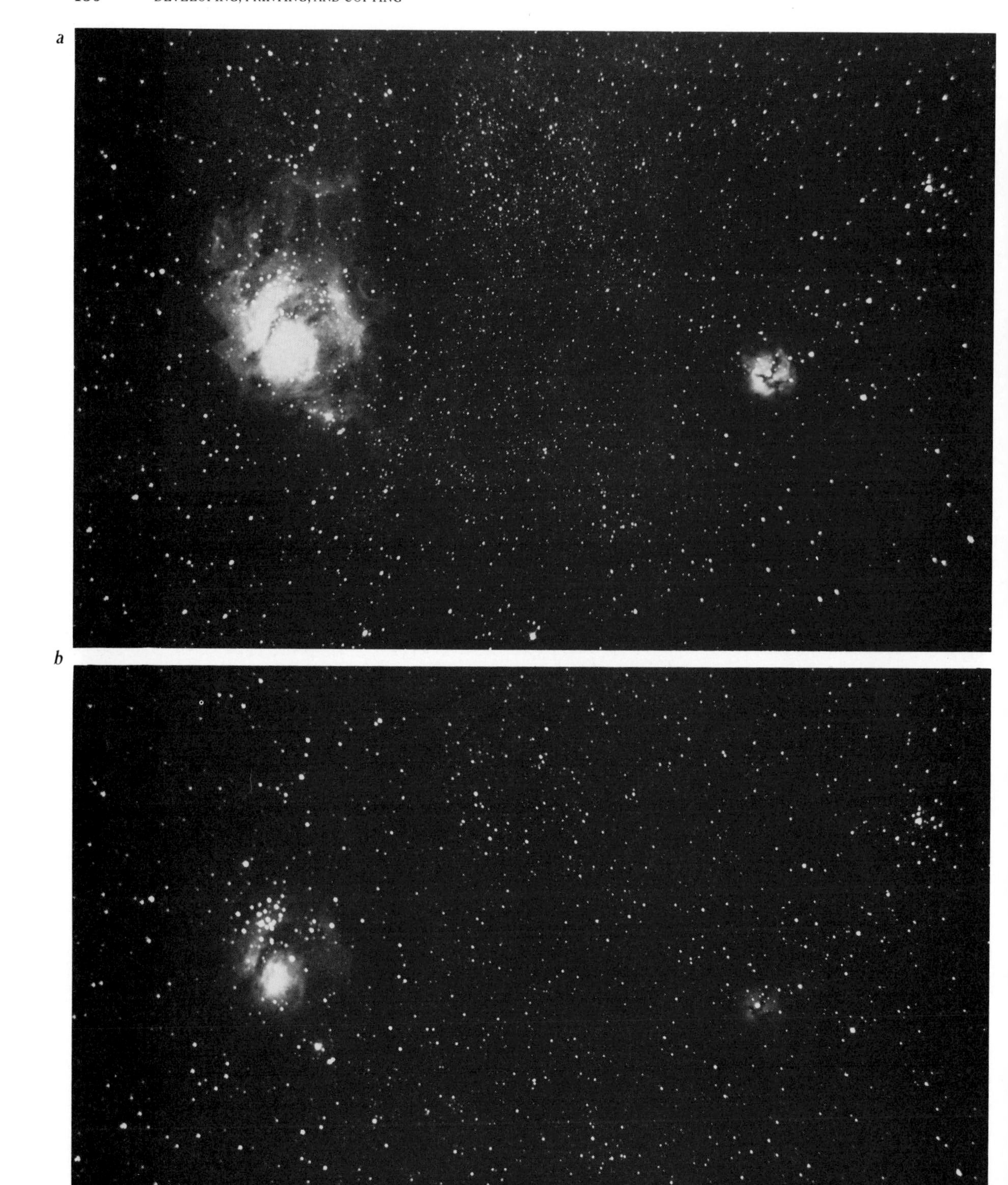
a
b

Fig. 10.5. The Lagoon and Trifid nebulae as photographed by Jim Baumgardt at the prime focus of a 15-cm (6-inch) f/4 Newtonian on gas-hypersensitized Technical Pan 2415. (a) 18 minutes through a red filter; (b) 28 minutes through a green filter; (c) 15 minutes through a blue filter. For the color picture formed by combining them, see Plate 10.2.

system of tricolor photography ('Simple and inexpensive tricolor photography', *Astronomy*, November 1983, pp. 51–4). The tricolor process is a way of making color pictures, not with conventional three-layer color film, but with three separate black-and-white negatives exposed through different colored filters. For instance, a typical star-field photograph might consist of a 24-minute exposure through a #25 red filter, a 22-minute exposure through a #44A blue filter, and a 32-minute exposure through a #11 green filter, all on gas-hypersensitized 2415. (The times are calculated to produce equal densities.)

A good 8×10 enlargement is then made from each negative, and the enlargements are retouched to eliminate dust spots. Then a precisely registered multiple exposure combining the three is made on color film. Each print is photographed through a filter of the same color as the one it was taken through, although the filters are now #25 (red), #38A (blue), and #58 (green), respectively. A suggested exposure ratio is 1 for red, 5.3 for blue, and 2.3 for green; that is, the blue exposure should be 5.3 times as long as the red one, and so forth, though considerable variation is possible. Fig. 10.5 shows a typical red-green-blue set, and Plate 10.2 shows the result of combining them. This is obviously a technique that requires much experimentation and careful record-keeping – but then, so does any other kind of astrophotography.

Appendix A
Sources of further information

Astrophotography is a combination of two popular hobbies and hence attracts two kinds of people, photographers who are new to astronomy and amateur astronomers who are new to photography.

One of the best ways to get more information about amateur astronomy is to read the major magazines – *Astronomy, Sky and Telescope,* and the journals of the principal amateur societies. You can find these magazines, as well as a wide selection of books, at public or university libraries.

You can often locate other amateur astronomers through a nearby planetarium, science museum, or college or university astronomy department.

Most booksellers can order any currently available astronomical book, even books they do not keep in stock, if you specify the author, title and publisher. Several specialist booksellers advertise in *Astronomy* and *Sky and Telescope.*

Getting information about photography is considerably easier, since any good camera shop or bookseller can supply books, and local clubs (often with darkrooms for members to use) are abundant. The single biggest source of information is the Eastman Kodak Company (addresses below), which publishes hundreds of books and pamphlets ranging from introductory to highly technical. In addition, I have listed a few of the many good books available from other publishers.

Organizations

American Association of Variable Star Observers, 25 Birch Street, Cambridge, Massachusetts 02138, USA.
Association of Lunar and Planetary Observers, c/o Mr H. D. Jamieson, PO Box 143, Heber Springs, Arkansas 72543, USA.
Astronomical Society of Southern Africa, PO Box 9, Observatory, CP 7935, South Africa.
British Astronomical Association, Burlington House, Piccadilly, London W1V 9AG, England. (Has members all over the world; publishes useful *Handbook* and *Journal.*)
Orange County Astronomers, c/o Mr John Sanford, 2215 Martha Avenue, Orange, California 92667, USA. (Holds astrophotography seminars and publishes proceedings.)
Royal Astronomical Society of Canada, 136 Dupont Street, Toronto, Ontario M5R 1V2, Canada.
Royal Astronomical Society of New Zealand, PO Box 3181, Wellington C1, New Zealand. (Plays a large role in coordinating amateur activities in the southern hemisphere.)
The Webb Society of Deep-Sky Observers, c/o Mr K. G. Jones, Wild Rose, Church Road, Winkfield, Windsor, Berkshire SL4 4SF, England. In the USA: c/o Mr R. J. Morales, 1440 S Marmora Avenue, Tucson, Arizona 85713.

Magazines

The Astrograph, Box 2283, Arlington, Virginia 22202, USA. (Devoted entirely to amateur astrophotography.)
Astronomy, 1027 N Seventh Street, Milwaukee, Wisconsin 53233, USA. (The same company publishes other astronomical magazines: *Odyssey* (for younger readers), *Telescope Making,* and *Deep Sky.*)
Astronomy Now, 193 Uxbridge Road, London W12 9RA, England. (Wide coverage).
Popular Astronomy, c/o Junior Astronomical Society, 36 Fairway, Keyworth, Nottingham NG12 5DU, England. (For beginners of all ages.)
Sky and Telescope, Sky Publishing Corporation, 49 Bay State Road, Cambridge, Massachusetts 02238-1290, USA.
Southern Astronomy, 116 Bronte Road, Bondi Junction, Sydney 2012, NSW, Australia.

Books – Introductory

BROWN, SAM. *All About Telescopes*, 5th edition, Barrington, New Jersey: Edmund Scientific Company, 1981. (Very good on basic observing techniques and how to find things in the sky; extremely readable and well-illustrated. Has good sections on optics and on telescope making; oriented toward home-built Newtonians.)

MENZEL, DONALD H. AND PASACHOFF, JAY M. *A Field Guide to the Stars and Planets*, 2nd edition, Boston: Houghton Mifflin, 1983. (Includes whole-sky maps of the constellations visible in the northern and southern hemispheres every month; an abridged version of Tirion's *Sky Atlas* showing stars to magnitude 7.0 and all deep-sky objects within reach of amateur instruments; moon maps; and other useful information.)

MOORE, PATRICK. *The Amateur Astronomer*, 11th edition, Cambridge, England: Cambridge University Press; New York: W. W. Norton, 1990. (A wide-ranging handbook that presupposes no prior knowledge. There are many other good books by the same author.)

PELTIER, LESLIE. *Leslie Peltier's Guide to the Stars*, Cambridge, England: Cambridge University Press, 1986. (Read this if you are unfamiliar with the night sky. It shows you how to learn the constellations and find celestial objects.)

RIDPATH, IAN. *Norton's 2000.0 Star Atlas and Reference Handbook*, London: Longman; New York: Halsted Press, 1989. (Completely updated version of Arthur P. Norton's classic. Highly recommended. The maps are especially easy to use because they look like the constellations in the sky.)

SCHOFIELD, JACK. (consulting editor). *The Darkroom Book*, New York: Ziff-Davis, 1981. (A guide to darkroom work, presupposing no prior knowledge.)

Books – Advanced

Astronomical Almanac for the Year 1991 (etc.), Washington: Government Printing Office; London: HM Stationery Office. (Positions and phenomena of the sun, moon, and planets. Libraries often classify this under government publications rather than astronomy.)

BURNHAM, ROBERT, JR. *Celestial Handbook*, revised edition, 3 vols, New York: Dover Publications, 1978–9. (Comprehensive information on all deep-sky objects within reach of amateur telescopes.)

Encyclopedia of Practical Photography, edited by the Eastman Kodak Company, 14 vols, Garden City, New York: Amphoto, 1977–9. (Thorough and authoritative; no longer fully up-to-date.)

Kodak Filters for Scientific and Technical Uses (Kodak Publication B-3), Rochester, New York: Eastman Kodak Company. (Spectral transmission curves for all Wratten-numbered filters.)

LANGFORD, MICHAEL. *The Master Guide to Photography*, New York: Knopf, 1982. (A thorough compendium; especially good on lenses and darkroom work. There are several other good books by the same author.)

SHERROD, P. CLAY, AND KOED, THOMAS L. *A Complete Manual of Amateur Astronomy*, Englewood Cliffs, New Jersey: Prentice-Hall, 1981. (Concentrates on how to make scientifically useful observations.)

TIRION, WIL. *Sky Atlas 2000.0*, Cambridge, England: Cambridge University Press, 1981. (The best atlas for interpreting star-field photographs. It shows stars to eight magnitude and numerous deep-sky objects, including diffuse nebulae.)

WALLIS, BRAD, AND PROVIN, ROBERT. *A Manual of Advanced Celestial Photography*, Cambridge, England: Cambridge University Press, 1988. (Comprehensive.)

Manufacturers

Broadhurst, Clarkson & Fuller (Fullerscopes), Telescope House, Farringdon Road, London EC1 M3JB. (Telescopes, accessories, and books. Sells its own line of telescopes and imports products of Meade, Celestron, Lumicon, and others. Large showroom.)
Celestron International, 2835 Columbia Street, Torrance, California 90503, USA. (Telescopes, especially Schmidt–Cassegrains.)
Société Kinoptik, division Clavé, 31–33, rue de Tlemcen, 75020 Paris, France. (Telescope parts of all kinds; premium-quality eyepieces.)
Eastman Kodak Company, Kodak Information Center, 343 State Street, Rochester, New York 14650, USA, phone 800-242-2424. In Britain, contact Customer Relations, Kodak Ltd, Hemel Hempstead, Hertfordshire HP1 1JU.
Meade Instruments Corporation, 1675 Toronto Way, Costa Mesa, California 92626, USA. (Telescopes of all kinds, eyepieces and other accessories.)
Tele-Vue, Inc., 20 Dexter Plaza, Pearl River, New York 10965, USA. (Plössl eyepieces of outstanding quality.)

Dealers

These are a few dealers who offer relatively hard-to-find items and who appear to be well established and reliable. Many other equally reputable suppliers of astrophotographic equipment can be located through their advertisements in astronomy magazines.
Edmund Scientific Company, 101 East Gloucester Pike, Barrington, New Jersey 08007, USA. (Telescopes, accessories, surplus lenses and other optical components.)
Eric Fishwick Ltd, Grange Valley, St Helens, Merseyside WA11 0XE, England. (Cameras, lenses, film and paper at discount prices.)
Freestyle Sales Company, 5124 Sunset Boulevard, Los Angeles, California 90027, USA. (Hard-to-find films; surplus and out-of-date film and paper at bargain prices.)
Lumicon, Inc., 2111 Research Drive, Livermore, California 94550, USA. (Deep-sky astrophotography supplies; gas-hypersensitization equipment; hypersensitized film.)
Orion Telescopes, PO Box 1158-T, Santa Cruz, California 95061, USA. (Telescopes and accessories of all kinds. Catalogue contains useful explanatory information for beginners.)
Astro-Optical Supplies (Amasco) Pty Ltd, 53 Hume Street, Crow's Nest 2065, NSW, Australia. (Telescopes and accessories.)
Sky Instruments, MPO Box 3164, Vancouver, British Columbia V6B 3Y6, Canada. (Telescopes and accessories.)
Spiratone, Inc., 135-06 Northern Boulevard, Flushing, New York 11354, USA. (Lenses, filters, slide duplicators, darkroom equipment, bulk loading equipment, and other photographic accessories, all at discount prices.)

Appendix B
Exposure tables

I present these exposure tables with a very important warning; *they are only approximations.* Astronomical exposures can never be calculated exactly, since unpredictable factors, especially variations in the transparency of the air, can throw them off by as much as two stops in either direction. These tables are meant only as a general guide to help you avoid making whole sets of exposures that are wildly off the mark.

These suggested exposures are based on the standard exposure formula,

$$t \text{ (seconds)} = f^2 / (A \times B)$$

where f is the f-ratio, A is the ISO (ASA) film speed, and B is a coefficient indicating the relative brightness of the object being photographed. The values of B used are shown on the charts; many, but not all, of them are taken from Robert Burnham's article 'Getting the correct exposure' (*Astronomy*, June 1981, pp. 51–5). However, I have added a few refinements.

To begin with, the exposure times have been corrected for reciprocity failure using the formula:

$$t_{\text{corrected}} = (t + 1)^{(1/p)} - 1$$

where p, the Schwarzschild exponent, is 0.7 (a typical value for commonly used films that have not been gas-hypersensitized).

Second, exposures between 1/2000 and 1 second have been rounded (on the basis of geometric rather than arithmetic means) to the nearest standard shutter speed setting. Exposures under 1/2000 second are indicated with the symbol '<<'; exposures of more than 300 seconds are marked '>>' since calculating them accurately would necessitate knowing the exact amount of reciprocity failure present (which, in turn, can depend on factors as capricious as the temperature and the humidity).

In referring to these tables, be sure to allow for any filters you may be using. Because of reciprocity failure, it is better not to apply the filter factor to the exposure time (as is usual in pictorial photography); instead, compute a corrected film speed as follows:

corrected speed = rated speed / filter factor

and then use the corrected value in referring to the tables.

Finally, *always* bracket your exposures at least one, and preferably two, stops to either side of the values given in the tables.

MOON – thin crescent
B = 10

f/	ISO (ASA) 32	ISO (ASA) 64	ISO (ASA) 100	ISO (ASA) 200	ISO (ASA) 400
2	1/60	1/125	1/250	1/500	1/1000
2.8	1/30	1/60	1/125	1/250	1/500
4	1/15	1/30	1/60	1/125	1/250
5.6	1/8	1/15	1/30	1/60	1/125
8	1/4	1/8	1/15	1/30	1/60
11	1/2	1/4	1/8	1/15	1/30
16	1	1/2	1/4	1/8	1/15
22	3	1	1/2	1/4	1/8
32	6	3	2	1/2	1/4
45	15	6	4	2	1/2
64	40	15	9	4	2
100	130	50	30	11	5
130	265	100	55	20	9
160	>>	180	100	40	15
200	>>	>>	180	70	30
250	>>	>>	>>	130	50
300	>>	>>	>>	210	80

MOON – wide crescent

Also: dimly lit features on terminator at any time

B = 20

f/	ISO (ASA) 32	ISO (ASA) 64	ISO (ASA) 100	ISO (ASA) 200	ISO (ASA) 400
2	1/125	1/250	1/500	1/1000	1/2000
2.8	1/60	1/125	1/250	1/500	1/1000
4	1/30	1/60	1/125	1/250	1/500
5.6	1/15	1/30	1/60	1/125	1/250
8	1/8	1/15	1/30	1/60	1/125
11	1/4	1/8	1/15	1/30	1/60
16	1/2	1/4	1/8	1/15	1/30
22	1	1/2	1/4	1/8	1/15
32	3	1	1/2	1/4	1/8
45	6	3	2	1/2	1/4
64	15	6	4	2	1/2
100	50	20	11	5	2
130	100	40	20	9	4
160	180	70	40	15	6
200	>>	130	70	30	11
250	>>	235	130	50	20
300	>>	>>	210	80	30

MOON – quarter phase

Also: brightly lit features on terminator at any time

B = 40

f/	ISO (ASA) 32	ISO (ASA) 64	ISO (ASA) 100	ISO (ASA) 200	ISO (ASA) 400
2	1/250	1/500	1/1000	1/2000	<<
2.8	1/125	1/250	1/500	1/1000	1/2000
4	1/60	1/125	1/250	1/500	1/1000
5.6	1/30	1/60	1/125	1/250	1/500
8	1/15	1/30	1/60	1/125	1/250
11	1/8	1/15	1/30	1/60	1/125
16	1/4	1/8	1/15	1/30	1/60
22	1/2	1/4	1/8	1/15	1/30
32	1	1/2	1/4	1/8	1/15
45	3	1	1/2	1/4	1/8
64	6	3	2	1/2	1/4
100	20	8	5	2	1/2
130	40	16	9	4	2
160	70	30	15	6	3
200	130	50	30	11	5
250	235	90	50	20	8
300	>>	150	80	30	13

MOON – gibbous

B = 80

f/	ISO (ASA) 32	ISO (ASA) 64	ISO (ASA) 100	ISO (ASA) 200	ISO (ASA) 400
2	1/500	1/1000	1/2000	<<	<<
2.8	1/250	1/500	1/1000	1/2000	<<
4	1/125	1/250	1/500	1/1000	1/2000
5.6	1/60	1/125	1/250	1/500	1/1000
8	1/30	1/60	1/125	1/250	1/500
11	1/15	1/30	1/60	1/125	1/250
16	1/8	1/15	1/30	1/60	1/125
22	1/4	1/8	1/15	1/30	1/60
32	1/2	1/4	1/8	1/15	1/30
45	1	1/2	1/4	1/8	1/15
64	3	1	1/2	1/4	1/8
100	8	4	2	1/2	1/4
130	16	7	4	2	1/2
160	30	11	6	3	1
200	50	20	11	5	2
250	90	35	20	8	4
300	150	60	30	13	6

MOON – full

B = 200

f/	ISO (ASA) 32	ISO (ASA) 64	ISO (ASA) 100	ISO (ASA) 200	ISO (ASA) 400
2	1/2000	<<	<<	<<	<<
2.8	1/1000	1/2000	1/2000	<<	<<
4	1/500	1/1000	1/1000	1/2000	<<
5.6	1/250	1/500	1/500	1/1000	1/2000
8	1/125	1/250	1/250	1/500	1/1000
11	1/60	1/125	1/125	1/250	1/500
16	1/30	1/60	1/60	1/125	1/250
22	1/15	1/30	1/30	1/60	1/125
32	1/8	1/15	1/15	1/30	1/60
45	1/4	1/8	1/8	1/15	1/30
64	1/2	1/4	1/4	1/8	1/15
100	3	1	1/2	1/4	1/8
130	5	2	1	1/2	1/4
160	9	4	2	1/2	1/4
200	15	6	4	2	1/2
250	25	11	6	3	1
300	45	17	10	4	2

MOON – partial eclipse
Exposing for penumbra (light portion) only
B = 50

f/	ISO (ASA) 32	ISO (ASA) 64	ISO (ASA) 100	ISO (ASA) 200	ISO (ASA) 400
2	1/500	1/1000	1/1000	1/2000	<<
2.8	1/250	1/500	1/500	1/1000	1/2000
4	1/125	1/250	1/250	1/500	1/1000
5.6	1/60	1/125	1/125	1/250	1/500
8	1/30	1/60	1/60	1/125	1/250
11	1/15	1/30	1/30	1/60	1/125
16	1/8	1/15	1/15	1/30	1/60
22	1/4	1/8	1/8	1/15	1/30
32	1/2	1/4	1/4	1/8	1/15
45	2	1/2	1/2	1/4	1/8
64	5	2	1	1/2	1/4
100	15	6	4	2	1/2
130	30	12	7	3	1
160	50	20	12	5	2
200	95	35	20	9	4
250	175	70	35	15	6
300	290	110	60	25	10

MOON – partial eclipse: umbra and penumbra together
Try a wide range of exposures – these are only suggestions.
B = 0.25

f/	ISO (ASA) 32	ISO (ASA) 64	ISO (ASA) 100	ISO (ASA) 200	ISO (ASA) 400
2	1/2	1/4	1/8	1/15	1/30
2.8	1	1/2	1/4	1/8	1/15
4	4	2	1/2	1/4	1/8
5.6	8	4	2	1/2	1/4
8	20	9	5	2	1/2
11	50	19	11	5	2
16	135	50	30	12	5
22	>>	125	65	25	11
32	>>	>>	185	70	30
45	>>	>>	>>	185	70
64	>>	>>	>>	>>	185
100	>>	>>	>>	>>	>>
130	>>	>>	>>	>>	>>
160	>>	>>	>>	>>	>>
200	>>	>>	>>	>>	>>
250	>>	>>	>>	>>	>>
300	>>	>>	>>	>>	>>

MOON – relatively light total eclipse
Try a wide range of exposures – these are only suggestions.
B = 0.05

f/	ISO (ASA) 32	ISO (ASA) 64	ISO (ASA) 100	ISO (ASA) 200	ISO (ASA) 400
2	5	2	1	1/2	1/4
2.8	11	5	3	1	1/2
4	30	11	6	3	1
5.6	70	25	15	6	3
8	180	70	40	15	6
11	>>	165	90	35	14
16	>>	>>	255	100	40
22	>>	>>	>>	235	90
32	>>	>>	>>	>>	255
45	>>	>>	>>	>>	>>
64	>>	>>	>>	>>	>>
100	>>	>>	>>	>>	>>
130	>>	>>	>>	>>	>>
160	>>	>>	>>	>>	>>
200	>>	>>	>>	>>	>>
250	>>	>>	>>	>>	>>
300	>>	>>	>>	>>	>>

MOON – relatively dark total eclipse
Try a wide range of exposures – these are only suggestions.
B = 0.005

f/	ISO (ASA) 32	ISO (ASA) 64	ISO (ASA) 100	ISO (ASA) 200	ISO (ASA) 400
2	95	35	20	9	4
2.8	240	90	50	20	8
4	>>	245	135	50	20
5.6	>>	>>	>>	130	50
8	>>	>>	>>	>>	135
11	>>	>>	>>	>>	>>
16	>>	>>	>>	>>	>>
22	>>	>>	>>	>>	>>
32	>>	>>	>>	>>	>>
45	>>	>>	>>	>>	>>
64	>>	>>	>>	>>	>>
100	>>	>>	>>	>>	>>
130	>>	>>	>>	>>	>>
160	>>	>>	>>	>>	>>
200	>>	>>	>>	>>	>>
250	>>	>>	>>	>>	>>
300	>>	>>	>>	>>	>>

SUN – full disk or partial eclipse
Through full aperture Solar-Skreen filter
B = 80

f/	ISO (ASA) 32	ISO (ASA) 64	ISO (ASA) 100	ISO (ASA) 200	ISO (ASA) 400
2	1/500	1/1000	1/2000	<<	<<
2.8	1/250	1/500	1/1000	1/2000	<<
4	1/125	1/250	1/500	1/1000	1/2000
5.6	1/60	1/125	1/250	1/500	1/1000
8	1/30	1/60	1/125	1/250	1/500
11	1/15	1/30	1/60	1/125	1/250
16	1/8	1/15	1/30	1/60	1/125
22	1/4	1/8	1/15	1/30	1/60
32	1/2	1/4	1/8	1/15	1/30
45	1	1/2	1/4	1/8	1/15
64	3	1	1/2	1/4	1/8
100	8	4	2	1/2	1/4
130	16	7	4	2	1/2
160	30	11	6	3	1
200	50	20	11	5	2
250	90	35	20	8	4
300	150	60	30	13	6

SUN – total eclipse: prominences
No filter
B = 50

f/	ISO (ASA) 32	ISO (ASA) 64	ISO (ASA) 100	ISO (ASA) 200	ISO (ASA) 400
2	1/500	1/1000	1/1000	1/2000	<<
2.8	1/250	1/500	1/500	1/1000	1/2000
4	1/125	1/250	1/250	1/500	1/1000
5.6	1/60	1/125	1/125	1/250	1/500
8	1/30	1/60	1/60	1/125	1/250
11	1/15	1/30	1/30	1/60	1/125
16	1/8	1/15	1/15	1/30	1/60
22	1/4	1/8	1/8	1/15	1/30
32	1/2	1/4	1/4	1/8	1/15
45	2	1/2	1/2	1/4	1/8
64	5	2	1	1/2	1/4
100	15	6	4	2	1/2
130	30	12	7	3	1
160	50	20	12	5	2
200	95	35	20	9	4
250	175	70	35	15	6
300	290	110	60	25	10

SUN – total eclipse: inner corona (3° field)
No filter
B = 5

f/	ISO (ASA) 32	ISO (ASA) 64	ISO (ASA) 100	ISO (ASA) 200	ISO (ASA) 400
2	1/30	1/60	1/125	1/250	1/500
2.8	1/15	1/30	1/60	1/125	1/250
4	1/8	1/15	1/30	1/60	1/125
5.6	1/4	1/8	1/15	1/30	1/60
8	1/2	1/4	1/8	1/15	1/30
11	1	1/2	1/4	1/8	1/15
16	3	1	1/2	1/4	1/8
22	6	3	1	1/2	1/4
32	15	6	4	2	1/2
45	40	15	9	4	2
64	100	40	20	9	4
100	>>	130	70	30	11
130	>>	265	145	55	20
160	>>	>>	255	100	40
200	>>	>>	>>	180	70
250	>>	>>	>>	>>	130
300	>>	>>	>>	>>	210

SUN – total eclipse: outer corona (10° field)
No filter
B = 1

f/	ISO (ASA) 32	ISO (ASA) 64	ISO (ASA) 100	ISO (ASA) 200	ISO (ASA) 400
2	1/8	1/15	1/30	1/60	1/125
2.8	1/4	1/8	1/15	1/30	1/60
4	1/2	1/4	1/8	1/15	1/30
5.6	1	1/2	1/4	1/8	1/15
8	4	2	1/2	1/4	1/8
11	8	3	2	1/2	1/4
16	20	9	5	2	1/2
22	50	19	11	5	2
32	135	50	30	12	5
45	>>	130	70	30	11
64	>>	>>	185	70	30
100	>>	>>	>>	245	95
130	>>	>>	>>	>>	195
160	>>	>>	>>	>>	>>
200	>>	>>	>>	>>	>>
250	>>	>>	>>	>>	>>
300	>>	>>	>>	>>	>>

VENUS

B = 400

f/	ISO (ASA) 32	ISO (ASA) 64	ISO (ASA) 100	ISO (ASA) 200	ISO (ASA) 400
2	<<	<<	<<	<<	<<
2.8	1/2000	<<	<<	<<	<<
4	1/1000	1/2000	1/2000	<<	<<
5.6	1/500	1/1000	1/1000	1/2000	<<
8	1/250	1/500	1/500	1/1000	1/2000
11	1/125	1/250	1/250	1/500	1/1000
16	1/60	1/125	1/125	1/250	1/500
22	1/30	1/60	1/60	1/125	1/250
32	1/15	1/30	1/30	1/60	1/125
45	1/8	1/15	1/15	1/30	1/60
64	1/4	1/8	1/8	1/15	1/30
100	1	1/2	1/4	1/8	1/15
130	2	1/2	1/2	1/4	1/8
160	4	2	1/2	1/4	1/8
200	6	3	2	1/2	1/4
250	11	5	3	1	1/2
300	17	7	4	2	1/2

JUPITER

B = 30

f/	ISO (ASA) 32	ISO (ASA) 64	ISO (ASA) 100	ISO (ASA) 200	ISO (ASA) 400
2	1/250	1/500	1/1000	1/2000	<<
2.8	1/125	1/250	1/500	1/1000	1/2000
4	1/60	1/125	1/250	1/500	1/1000
5.6	1/30	1/60	1/125	1/250	1/500
8	1/15	1/30	1/60	1/125	1/250
11	1/8	1/15	1/30	1/60	1/125
16	1/4	1/8	1/15	1/30	1/60
22	1/2	1/4	1/8	1/15	1/30
32	2	1/2	1/4	1/8	1/15
45	4	2	1/2	1/4	1/8
64	9	4	2	1/2	1/4
100	30	12	7	3	1
130	60	25	13	6	2
160	105	40	20	9	4
200	190	75	40	16	7
250	>>	135	75	30	12
300	>>	225	120	50	19

MERCURY or MARS

B = 60

f/	ISO (ASA) 32	ISO (ASA) 64	ISO (ASA) 100	ISO (ASA) 200	ISO (ASA) 400
2	1/500	1/1000	1/2000	<<	<<
2.8	1/250	1/500	1/1000	1/2000	<<
4	1/125	1/250	1/500	1/1000	1/2000
5.6	1/60	1/125	1/250	1/500	1/1000
8	1/30	1/60	1/125	1/250	1/500
11	1/15	1/30	1/60	1/125	1/250
16	1/8	1/15	1/30	1/60	1/125
22	1/4	1/8	1/15	1/30	1/60
32	1/2	1/4	1/8	1/15	1/30
45	2	1/2	1/4	1/8	1/15
64	4	2	1/2	1/4	1/8
100	12	5	3	1	1/2
130	25	10	6	2	1
160	40	16	9	4	2
200	75	30	16	7	3
250	135	55	30	12	5
300	225	85	50	19	8

SATURN

B = 10

f/	ISO (ASA) 32	ISO (ASA) 64	ISO (ASA) 100	ISO (ASA) 200	ISO (ASA) 400
2	1/60	1/125	1/250	1/500	1/1000
2.8	1/30	1/60	1/125	1/250	1/500
4	1/15	1/30	1/60	1/125	1/250
5.6	1/8	1/15	1/30	1/60	1/125
8	1/4	1/8	1/15	1/30	1/60
11	1/2	1/4	1/8	1/15	1/30
16	1	1/2	1/4	1/8	1/15
22	3	1	1/2	1/4	1/8
32	6	3	2	1/2	1/4
45	15	6	4	2	1/2
64	40	15	9	4	2
100	130	50	30	11	5
130	265	100	55	20	9
160	>>	180	100	40	15
200	>>	>>	180	70	30
250	>>	>>	>>	130	50
300	>>	>>	>>	210	80

Appendix C
Computer programs

The following computer programs, in BASIC, perform some calculations often needed in astrophotography: focal length and field of view of various optical configurations, image size, exposure, and equivalent black-and-white film development times at different temperatures.

I have gone to some effort to write programs that should run in even the most restrictive versions of BASIC; they are not as short or as straightforward as they could be if I had used Boolean expressions or compound IF statements, which are not supported on all computers. The one nonstandard statement that I have used is CLS, which (in Microsoft BASIC) clears the screen and homes the cursor. Wherever you see a CLS in these programs, substitute whatever statement performs the same function on your computer; if there is none, substitute PRINT. (Sinclair ZX81 and TS1000 users will have to restructure the exposure program to avoid using READ and DATA statements.)

The display in all cases fits on a 24-row by 40-column screen, and memory requirements are quite small (less than 3K for the exposure program, which is the largest).

In giving their numerical answers, these programs may disagree slightly with the tables elsewhere in the book, or even with each other, for several reasons. The exposure program customizes the reciprocity failure correction for the type of film used, whereas the exposure tables in Appendix B use a compromise value to cover a wide range of films. The image size program calculates image size with the tangent function, while the image sizes given by the optical configuration program are obtained with a simplified formula. Finally, all computers introduce rounding errors into their calculations; the amount of error depends on the exact order in which the calculations are done. In all these cases, the error is insignificant because the numbers from which the calculations are done are approximate in the first place.

Optical configuration program

This program calculates the effective focal length, *f*-ratio, field of view and sizes of the images of the moon and Jupiter for a prime focus, afocal, positive projection, negative projection, or compression system. For negative projection and compression, the required back focus (measured from the lens) is also computed.

```
10 CLS
20 PRINT "Optical configuration program"
30 PRINT "Michael A. Covington     1983"
40 PRINT
50 PRINT "Which optical configuration will"
60 PRINT "you be using?"
70 PRINT
80 PRINT "  1 -- Prime focus"
90 PRINT "  2 -- Afocal"
100 PRINT "  3 -- Positive projection"
110 PRINT "  4 -- Negative projection"
120 PRINT "  5 -- Compression"
130 PRINT
140 PRINT "Your choice (1 to 5)";
150 INPUT C
160 IF C<1 THEN 140
170 IF C>5 THEN 140
180 PRINT
190 PRINT "Telescope diameter (mm)";
200 INPUT A
210 PRINT
220 PRINT "Telescope focal length (mm)";
230 INPUT F
240 ON C GOTO 250,280,370,370,370
250 REM --- Prime focus ---
260 LET M=1
270 GOTO 520
280 REM --- Afocal ---
290 PRINT
300 PRINT "Camera lens focal length (mm)";
310 INPUT F1
320 PRINT
330 PRINT "Eyepiece focal length (mm)";
340 INPUT F2
350 LET M = F1/F2
360 GOTO 520
370 REM  --- Projection systems ---
380 LET N$=" projection "
390 IF C<>4 THEN 410
400 LET N$=" compressor "
410 PRINT
420 PRINT "Focal length of";N$;"lens"
430 PRINT "(mm)";
440 INPUT F2
450 IF C<>4 THEN 470
460 LET F2 = -ABS(F2)
470 PRINT
480 PRINT "Distance from";N$;"lens to film"
490 PRINT "(mm)";
500 INPUT S2
510 LET M=(S2-F2)/F2
520 REM  --- Calculations common to all configs. ---
530 CLS
```

```
540 PRINT
550 LET F9=F*ABS(M)
560 PRINT "System focal length: ";F9;" mm"
570 PRINT
580 PRINT "System f-ratio:  f/";F9/A
590 IF M>0 THEN 630
600 PRINT
610 PRINT "Back focus required: ";ABS(S2/M);" mm"
620 PRINT "  (measured from";N$;"lens)"
630 PRINT
640 REM --- Field of view calculations ---
650 REM  W1 and W2 are dimensions of the film
660 REM   (in mm); change if not using 35-mm format
670 LET W1 = 24
680 LET W2 = 36
690 REM  The following works correctly whether
700 REM   your computer's trig functions are
710 REM   in degrees or radians
720 LET Q = 45/ATN(1)
730 LET T1 = Q*2*ATN((W1/2)/F9)
740 LET T2 = Q*2*ATN((W2/2)/F9)
750 LET U$ = " degrees "
760 IF T1>1 THEN 840
770 LET T1 = T1 * 60
780 LET T2 = T2 * 60
790 LET U$ = " arc-minutes "
800 IF T1>1 THEN 840
810 LET T1 = T1 * 60
820 LET T2 = T2 * 60
830 LET U$ = " arc-seconds "
840 PRINT
850 PRINT "Field of view"
860 PRINT " (assuming no vignetting):"
870 PRINT
880 PRINT T1;" x ";T2;U$
890 LET J = F9 / 5000
900 PRINT
910 PRINT
920 PRINT "Jupiter image size:"
930 PRINT
940 PRINT "  On film:              ";J;" mm"
950 PRINT "  On 15x enlargement:";J*15;" mm"
960 PRINT
970 PRINT
980 PRINT
990 PRINT "Do you want to do another?"
1000 INPUT R$
1010 IF R$="y" THEN 10
1020 IF R$="Y" THEN 10
1030 IF R$="yes" THEN 10
1040 IF R$="YES" THEN 10
1050 END
```

Image size program

Given the angular size of any celestial object and the focal length of the system used to photograph it, this program calculates the image size on the film and on an enlargement of a given size.

```
10 CLS
20 PRINT "Image size program"
30 PRINT "Michael A. Covington  1983"
40 PRINT
50 PRINT "System focal length (mm)";
60 INPUT F
70 PRINT
80 PRINT "Enlargement factor";
90 INPUT E
100 PRINT
110 PRINT "What units will you be using for"
120 PRINT "the size of the object photographed?"
130 PRINT
140 PRINT "  1. Degrees"
150 PRINT "  2. Arc-minutes"
160 PRINT "  3. Arc-seconds"
170 PRINT
180 PRINT "Your choice (1 to 3)";
190 INPUT C
200 IF C<1 THEN 180
210 IF C>3 THEN 180
220 PRINT
230 PRINT "Size of object photographed";
240 INPUT T
250 REM -- Convert size to degrees --
260 IF C=1 THEN 310
270 LET T=T/60
280 IF C=2 THEN 310
290 LET T=T/60
300 REM -- Convert to computer's units --
310 REM  The following works whether
320 REM   your computer's trig functions
330 REM   use degrees or radians.
340 LET Q = 45 / ATN(1)
350 LET T=T/Q
360 REM -- Find image size --
370 LET W = 2*F*TAN(T/2)
380 PRINT
390 PRINT "Image size:"
400 PRINT "  On film, ";W;" mm"
410 PRINT "  On ";E;"X enlargement, ";W*E;" mm"
420 PRINT
430 PRINT "Do you wish to do another";
440 INPUT R$
450 IF R$="y" THEN 10
460 IF R$="Y" THEN 10
470 IF R$="yes" THEN 10
480 IF R$="YES" THEN 10
490 END
```

Exposure program

This program calculates exposures for celestial objects using formulae similar to those used to generate the tables in Appendix B. You supply the *f*-ratio, film speed, type of reciprocity correction desired, and filter factor (if any); the program displays exposure times for a number of celestial objects.

```
10 CLS
20 PRINT "Astrophotographic exposure program"
30 PRINT "Michael A. Covington          1983"
40 PRINT
50 DIM S(12), S$(12), P(3)
60 REM --- List of standard shutter speeds ---
70 DATA 0.0005,"1/2000",0.001,"1/1000"
80 DATA 0.002,"1/500",0.004,"1/250"
90 DATA 0.008,"1/125",0.016667,"1/60"
100 DATA 0.033333,"1/30",0.066667,"1/15"
110 DATA 0.125,"1/8",0.25,"1/4"
120 DATA 0.5,"1/2",1," 1"
130 REM  Shutter speeds will be compared
140 REM   logarithmically.  Base 10 or base E
150 REM   logarithms are equally OK.
160 FOR I=1 TO 12
170   READ S(I),S$(I)
180   LET S(I)=LOG(S(I))
190 NEXT I
200 REM --- List of objects and B values ---
210 DATA "Moon (thin crescent)...  ",10
220 DATA "Moon (wide crescent)...  ",20
230 DATA "Moon (quarter).........  ",40
240 DATA "Moon (gibbous).........  ",80
250 DATA "Moon (full)............  ",200
260 DATA "Mercury................  ",60
270 DATA "Venus..................  ",400
280 DATA "Mars...................  ",60
290 DATA "Jupiter................  ",30
300 DATA "Saturn.................  ",10
310 DATA "END"
320 REM "END" in preceding line must be all capital
     letters
330 REM -- Obtain parameters for this run ---
340 PRINT
350 PRINT "System f/ratio:  f/";
360 INPUT F
370 PRINT
```

```
380 PRINT "ISO (ASA) film speed";
390 INPUT A
400 PRINT
410 PRINT "Type of reciprocity correction desired:"
420 PRINT
430 PRINT "    1.  Typical fast film (e.g., Tri-X)"
440 PRINT "    2.  Typical slow film (e.g., 2415)"
450 PRINT "    3.  No correction (hypersensitized"
460 PRINT "                       or Spectroscopic film)"
470 PRINT
480 PRINT "Your choice (1, 2, or 3)";
490 INPUT C
500 IF C<1 THEN 480
510 IF C>3 THEN 480
520 REM  P(C) will be Schwarzschild exponent...
530 LET P(1) = .65
540 LET P(2) = .8
550 LET P(3) = 1
560 PRINT
570 PRINT "Will you be using a filter";
580 INPUT R$
590 LET F1 = 1
600 IF R$="n" THEN 670
610 IF R$="N" THEN 670
620 IF R$="no" THEN 670
630 IF R$="No" THEN 670
640 PRINT
650 PRINT "Filter factor";
660 INPUT F1
670 REM --- Set up to display exposure table ---
680 CLS
690 PRINT "Film speed: ISO ";A
700 PRINT "F-ratio:      f/ ";F
710 PRINT "Filter factor = ";F1
720 PRINT
730 REM --- Apply filter factor ---
740 LET A = A/F1
750 REM --- Main loop ---
760 READ N$
770 IF N$="END" THEN 1060
780 PRINT N$;
790 READ B
800 REM  Exposure formula
810 LET T = (F^2)/(A*B)
820 REM  Reciprocity correction
830 LET T = (T+1) ^ (1/P(C)) - 1
840 REM  If over 5 minutes, say so
850 IF T<300 THEN 890
860 PRINT "Over 5 min."
870 GOTO 750
880 REM  If over 1.4 seconds, round and print
890 IF T<1.4 THEN 960
900 PRINT INT(T+.5);" sec."
910 GOTO 750
920 REM  If under 1/2000, say so
930 IF T>.0005 THEN 960
940 PRINT "Under 1/2000 sec."
950 GOTO 750
960 REM  Find nearest std. shutter speed
970 REM    (on logarithmic scale)
980 LET L = LOG(T)
990 LET S7 = 1
1000 FOR S9 = 2 TO 12
1010    IF ABS(S(S9)-L) > ABS(S(S7)-L) THEN 1030
1020    LET S7 = S9
1030 NEXT S9
1040 PRINT S$(S7);" sec."
1050 GOTO 750
1060 REM --- Termination ---
1070 PRINT
1080 PRINT "Do you wish to do another";
1090 INPUT R$
1100 IF R$="YES" THEN 1160
1110 IF R$="yes" THEN 1160
1120 IF R$="Y" THEN 1160
1130 IF R$="y" THEN 1160
1140 END
1150 GOTO 1190
1160 CLS
1170 RESTORE
1180 GOTO 60
1190 END
```

Development time program

This program converts black-and-white film development times from one temperature to another. The resulting values are only approximations and are most accurate with general-purpose developers such as HC-110 and D-76.

```
10 CLS
20 PRINT "Development time program"
30 PRINT "Michael A. Covington  1983"
40 PRINT
50 PRINT "This program calculates development"
60 PRINT "times for black-and-white film"
70 PRINT "processed at temperatures other than"
80 PRINT "specified in the instructions."
90 PRINT
100 PRINT "Select temperature units to be used:"
110 PRINT
120 PRINT "   1  Centigrade"
130 PRINT "   2  Fahrenheit"
140 PRINT
150 PRINT "Type 1 or 2 --";
160 INPUT R$
170 LET K = -.08
180 LET U$ = "C"
190 IF R$ = "1" THEN 230
200 LET K = -.045
210 LET U$ = "F"
220 IF R$ <> "2" THEN 150
230 PRINT
240 PRINT "Originally specified time (minutes)?"
250 INPUT Z1
260 PRINT
270 PRINT "Originally specified temperature (";U$;")?"
280 INPUT T1
290 PRINT
300 PRINT "New temperature (";U$;")?"
310 INPUT T2
320 PRINT
330 IF ABS(K*(T1-T2)) < .5 THEN 370
340 PRINT "Temperature difference is too large"
350 PRINT "for accurate calculation."
360 GOTO 470
370 LET Z2 = Z1 * EXP(K*(T2-T1))
380 PRINT "New development time:"
390 PRINT Z2;" minutes."
400 IF Z2>5 THEN 470
410 PRINT
420 PRINT "Note: Development times less than"
430 PRINT "5 minutes produce poor uniformity"
440 PRINT "and this program may not calculate"
450 PRINT "them accurately."
460 PRINT
470 PRINT
480 PRINT "Do you want to do another?"
490 INPUT R$
500 IF R$="YES" THEN 10
510 IF R$="yes" THEN 10
520 IF R$="Y" THEN 10
530 IF R$="y" THEN 10
540 END
```

Appendix D
Plans for an electronic guider

The circuits

Chapter 7 described the use of an electronic drive corrector, of which many models are commercially available. This appendix gives plans for a low-priced home-built drive corrector, plus a quartz frequency standard to aid in adjusting the drive rate, and a power supply to power the drive corrector from the 120- or 240-volt line.

The circuit for the drive corrector is shown in Fig. D.1. It can be built to deliver either 120 or 240 volts at a center frequency of either 50 or 60 Hz. Powered by a 12-volt car battery or the like, it draws about 0.7 ampere under light load and can deliver 10 watts of output without difficulty, enough to power most amateur telescopes up to about 25 cm (10 inches) aperture. The operating frequency is adjustable, and instantaneous corrections are made with the 'fast' and 'slow' push-button switches.

The quartz frequency standard (Fig. D.2) provides a flashing-LED indication of how close the operating frequency is to 50 or 60 Hertz; it is meant to be built on the same circuit board as the drive corrector itself.

The mains power supply (Fig. D.3) provides a ripple-free 12-volt supply for the drive corrector and frequency standard. If there is ripple (hum) on the 12-volt supply line, the drive corrector tends to lock onto it and synchronize with the ripple frequency, making fine control of drive speed impossible; this circuit uses IC voltage regulation to keep ripple well below the lowest level at which it could have any effect.

Fig. D.1. *Circuit diagram of the electronic drive corrector.*

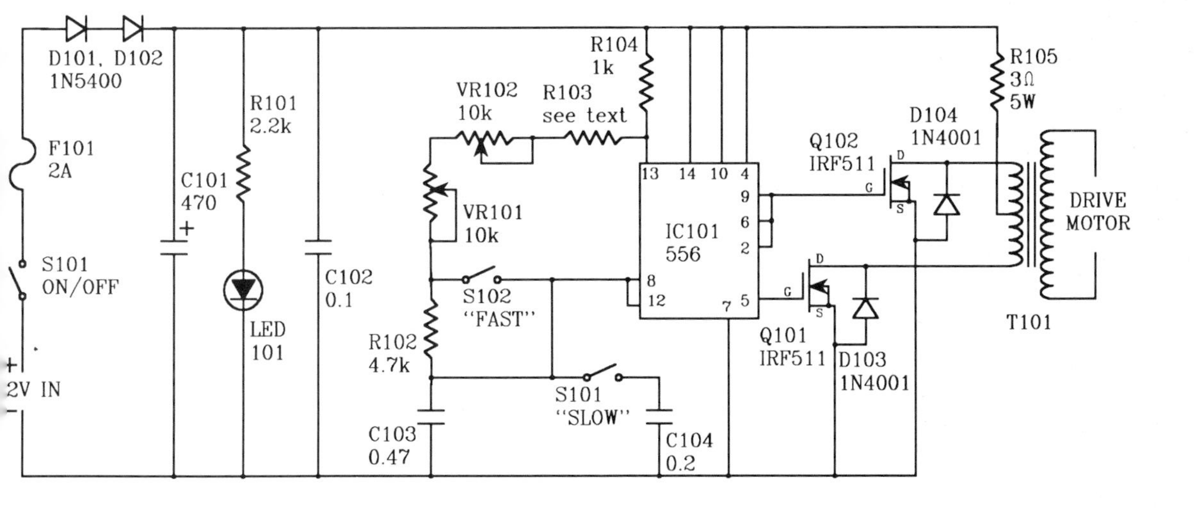

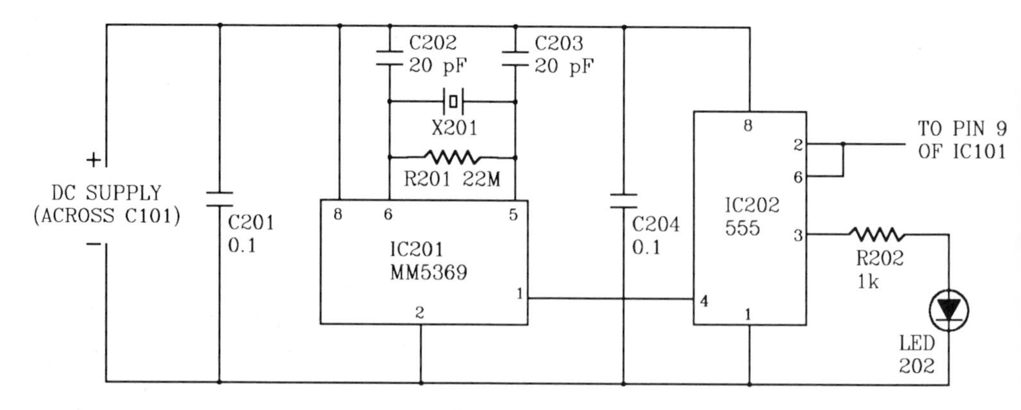

Fig. D.2. Circuit of the quartz frequency standard.

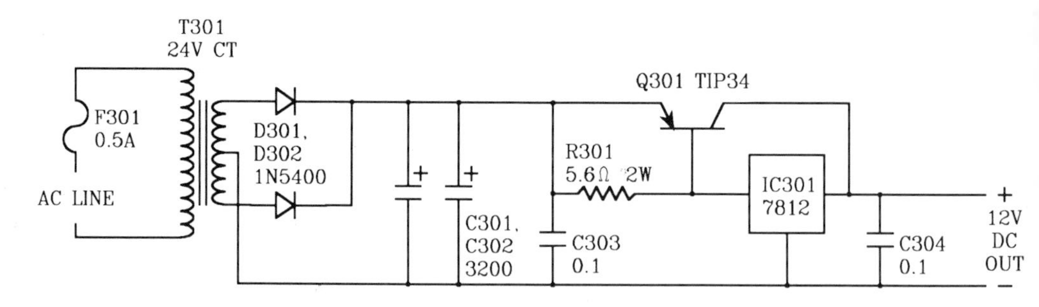

Fig. D.3. The mains power supply for the electronic drive corrector.

Parts list

In the following combined list, parts in the drive corrector itself are numbered from 101; those in the quartz frequency standard, from 201; and those in the mains power supply, from 301.

Resistors

(Higher wattages can always be substituted. 10% or better tolerance is adequate.)
R101 – 2200 ohms, ⅛ watt.
R102 – 4700 ohms, ⅛ watt.
R103 – 10 000 ohms (for 60 Hz), or 15 000 ohms (for 50 Hz), ⅛ watt.
R104, R202 – 1000 ohms, ¼ watt.
R105 – 3 ohms, 5 watts. (If not available, use three 10-ohm, 2-watt resistors in parallel.) This resistor gives off quite a bit of heat and should not be in contact with other components.
R201 – 20 or 22 megohms, ⅛ watt. (If not available, use two 10-megohm resistors in series.)
R301 – 5.6 ohms, 2 watts.
VR101 – 10 000-ohm linear taper potentiometer.
VR102 – 10 000-ohm trimmer potentiometer (to be mounted on the circuit board).

Capacitors

(In most cases, tolerances are so loose that I have specified a range of values.)
C101 – 470 to 3200 microfarads, 25 to 50 volts, electrolytic.
C102, C201, C204, C303, C304 – 0.1 to 0.47 microfarad, 50 volts or higher, ceramic disk, mylar, polyester, or paper. Each of these should be placed

Fig. D.4. A substitute for the power MOSFET used in the drive corrector.

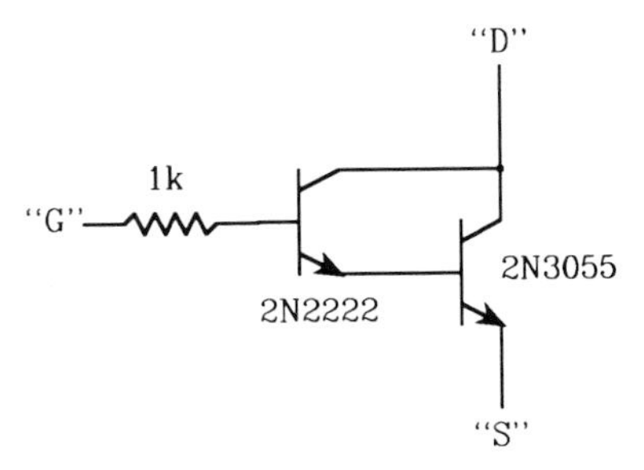

as close as possible to the IC with which it is associated.
C103 – 0.47 microfarad, 10% or better tolerance, 50 volts or higher, mylar or polyester (for good temperature stability).
C104 – 0.2 microfarad, 20% or better tolerance, 50 volts or higher.
C202, C203 – 20 to 50 picofarads, 50 volts or higher.
C301, C302 – 3200 microfarads or higher, 25 to 50 volts.

Semiconductors

LED101, LED201 – general-purpose red light-emitting diode. (Red is less disturbing to night vision than other colors.)
D101, D102, D301, D302 – type 1N5400 or equivalent silicon rectifiers rated at 2 or more amperes forward current, 50 or more peak reverse volts.
D103, D104 – type 1N4001 or equivalent silicon rectifiers rated at 1 or more amperes forward current, 50 or more peak reverse volts.
IC101 – type 556 dual timer.
IC201 – crystal oscillator/divider, type MM5369AA (for 60 Hz) or MM5369EYR (for 50 Hz). (Versions of the MM5369 have been manufactured for several different combinations of frequencies; you want the one that takes a 3.58-MHz crystal and delivers 50 or 60 Hz as the case may be.)
IC202 – type 555 timer.
IC301 – type 7812 or equivalent 12-volt positive voltage regulator. (In this circuit, the regulator carries about 0.12 ampere of current and dissipates about 1 watt. Under these circumstances the 7812 does not require a heat sink.)
Q101, Q102 – type IRF-511 or equivalent power MOSFET rated at 3 or more amperes forward current and no more than 0.6 ohm 'on' resistance. (If not available, see Fig. D.4 for a substitute.) These transistors give off some heat in use, but a heat sink is not required.
Q301 – PNP silicon power transistor rated at 4 or more amperes continuous collector current and at least 10 watts continuous dissipation. (Many types are suitable – among them MJE34, TIP34, MJE2955, TIP2955, TIP42, and ECG-153.) *Mount this transistor on a heat sink.*

Crystal

X201 – 3.58-MHz quartz crystal (as used in American color TV receivers – actually 3.579545 MHz).

Transformers

T101 – 18-volt, 2-ampere, center-tapped power transformer. (The primary voltage should be the voltage required by the telescope drive, 120 or 240 volts.)
T301 – 24-volt or 25.2-volt, 2-ampere, center-tapped power transformer. (The primary voltage should be the local line voltage.)

Fuses

F101 – 2-ampere normal-acting fuse.
F301 – ½-ampere slow-blow fuse.

Switches

S101 – SPST toggle switch rated to carry 2 amperes at 12 volts DC.
S102, S103 – miniature normally-open SPST push-button switches.

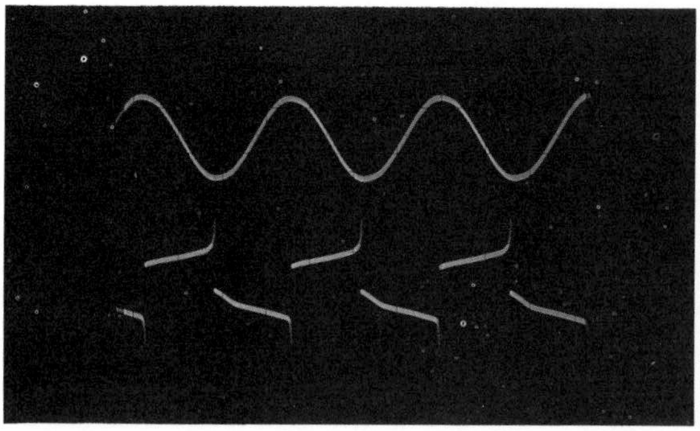

Fig. D.5. Oscilloscope traces of the 120-volt, 60-Hz line voltage (top) and the output of the drive corrector (bottom).

How it works

The first part of the drive corrector circuit conditions the incoming DC power. Diodes D101 and D102 prevent damage if the supply is connected backward, and also drop the operating voltage to about 11 volts. (If you routinely use a power supply that runs higher than 12 volts, you may want to add more diodes, one for every 0.6 volt of additional drop desired.) C101 and C102 smooth out fluctuations that arise because the output stage draws power in short bursts rather than continuously. LED101 is, of course, the 'power on' indicator; if it is too bright, increase R101 to 4700 ohms.

IC101, a type 556 dual timer, generates the 50- or 60-Hz signal. Half of it functions as an astable multivibrator whose frequency is determined by a network of resistors and capacitors; the other half is used as an inverter, so that two complementary outputs are obtained. The 'fast' switch bypasses a resistor, and the 'slow' switch adds an extra capacitor.

The output stage is a push-pull amplifier in which two power MOSFETs drive the low-voltage side of an 18-volt center-tapped power transformer. R105 limits the saturation current through the transformer. The 18-volt transformer goes into saturation on each cycle; operating in saturation mode does waste some power, but it ensures that the output voltage is relatively independent of the supply voltage and cannot go high enough to damage the drive motor.

As Fig. D.5 shows, the output of the drive corrector is a distorted square wave. Ordinary AC voltmeters give correct readings only when the input is a sine wave; hence attempts to measure the output voltage may give misleading results. A better test is to use the output to light a 7-watt bulb, and compare its brightness to that obtained with mains power. Note that drive motors are quite insensitive to voltage variation.

The quartz frequency standard uses a type MM5369 oscillator/divider IC to obtain a 50- or 60-Hz reference signal, which is then compared with the drive corrector frequency by means of a 555 timer used as a gate. The result is that LED201 flashes at a rate equal to the difference between the two frequencies (for example, 2 Hz if the drive corrector is running at 58 Hz and the reference signal is 60 Hz). To 'tune in' on the correct frequency, simply adjust VR101 to make the LED flash as slowly as possible.

Finally, the ripple-free mains power supply uses a 7812 voltage regulator assisted by a power transistor. Note that this circuit is not overload protected; damage may result if the output is shorted.

Construction, adjustment, and use

Figs. D.6 and D.7 show my version of the drive corrector; yours may, of course, look quite different.

Construction is not particularly tricky. Q101 and Q102 are somewhat sensitive to static electricity and should be left in their original packaging until you are ready to solder them in. After you get everything working, it may be a good idea to spray all the components with insulating varnish to protect them from dew.

If you have built the quartz frequency standard, frequency adjustment is simply a matter of setting the main variable resistor (VR101) to the middle of its scale and adjusting the trimmer (VR102) to make LED201 flash as slowly as possible. If you do not have the quartz frequency standard, you'll have to do the adjusting by tracking a star with your telescope;

start with both VR101 and VR102 in the middle of their ranges, and adjust VR102 until the drive rate is correct. An alternative is to use the drive corrector to power a (motor-driven) electric clock and adjust until it runs at the right speed.

In use, the drive corrector should be given a few minutes to reach operating temperature before fine adjustment of speed is attempted; temperature changes are the main cause of drift, and since R105, Q101, and Q102 give off heat, some drift is inevitable.

With the quartz frequency standard, 'tuning in' to make the LED flash as slowly as possible gives you the solar rate (one revolution in 24 hours, suitable for planetary photography). To get the sidereal rate for deep-sky photography, tune in on the solar rate, then increase the speed until the LED flashes about once every six seconds. To get a good approximation to the lunar rate, tune in and then lower the speed until you get approximately two flashes per second.

When operating very close to the solar rate, the drive corrector may lock in on oscillations from the quartz frequency standard, causing the LED to stop flashing altogether. This may not be undesirable; if you want to prevent it, keep the components of the quartz frequency standard well away from those of the drive corrector on the circuit board. Also, if operated indoors or close to power lines, the drive corrector may pick up the stray electrical field and synchronize with the power line frequency, especially if the cable connecting the push-button controls to the main unit is not shielded.

Note that, because the output transformer operates in saturation mode, the drive corrector draws nearly as much current without a load as with one. To conserve battery power, turn the clock drive on and off with S101 rather than at the motor.

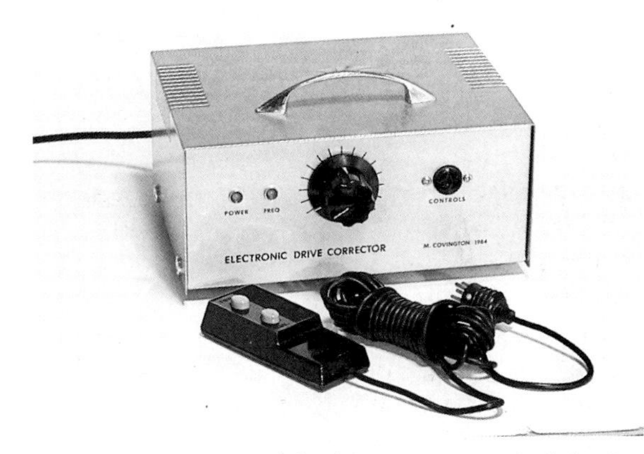

Fig. D.6. *Front view of the drive corrector as built by the author. The plug-in remote control is from a slide projector.*

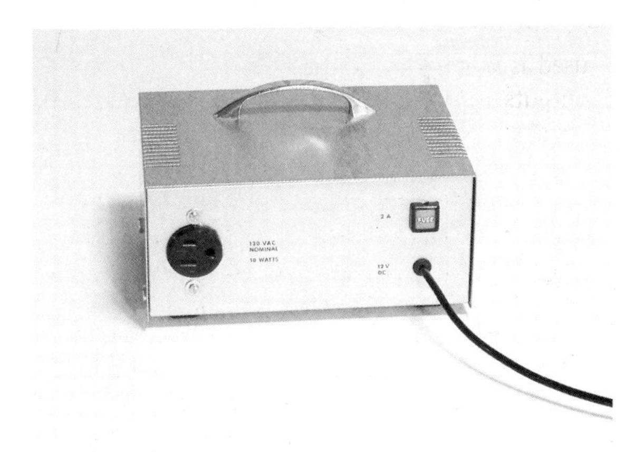

Fig. D.7. *Rear view of the drive corrector, showing the socket for connecting the clock drive.*

Appendix E
Film data sheets

The following information is reprinted from data sheets by courtesy of Fuji Photo Film USA and Eastman Kodak Company. The materials covered are:

- Fujichrome 100 and 400 Professional D Films
- Kodak Ektachrome 200 Professional Film
- Kodak Ektachrome 100 HC Film
- Kodak T-Max 100 and 400 Professional Films
- Kodak T-Max P3200 Professional Film
- Kodak Technical Pan Films (2415, 4415, 6415)

For complete, up-to-date information you should obtain current data sheets from the manufacturers.

Fujichrome 100 and 400 Professional D Films

FUJICHROME 100 Professional D [RDP] ISO 100/21°, Daylight Type

• SPECTRAL SENSITIVITY CURVES

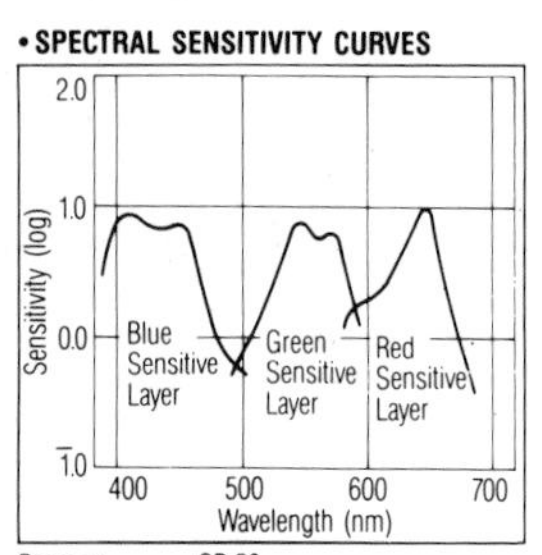

Process: CR-56
Densitometry: Status A
Density: 1.0 above minimum density
Sensitivity equals reciprocal of exposure (ergs/cm²) required to produce specified density

• SPECTRAL DYE DENSITY CURVES

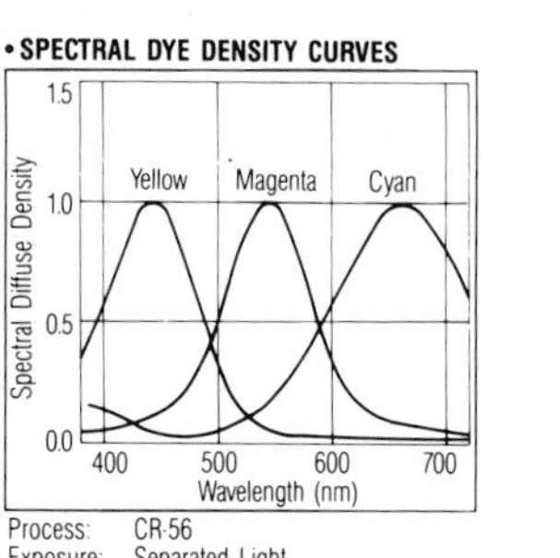
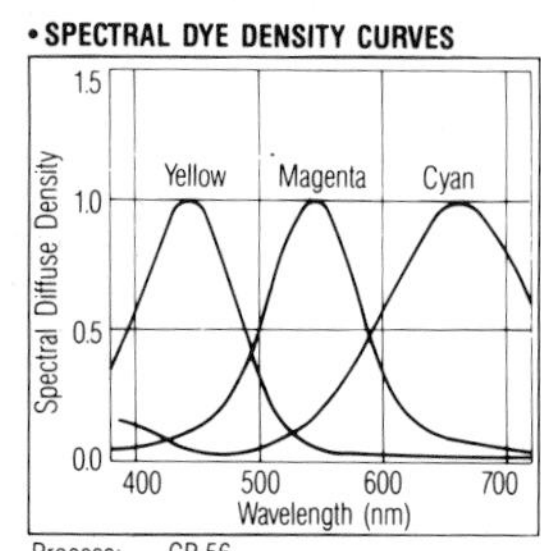
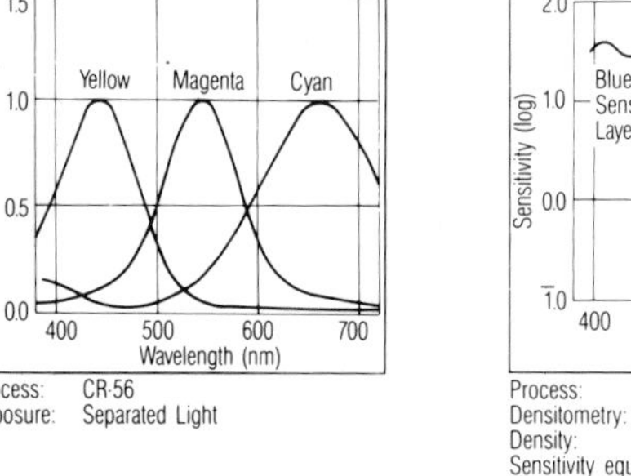

Process: CR-56
Exposure: Separated Light

• CHARACTERISTIC CURVES

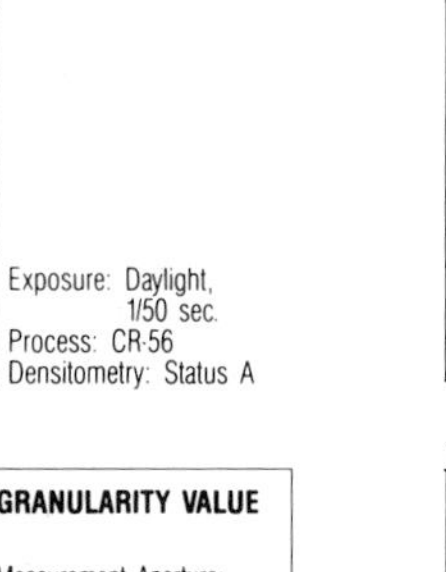

Exposure: Daylight, 1/50 sec.
Process: CR-56
Densitometry: Status A

• MTF CURVE

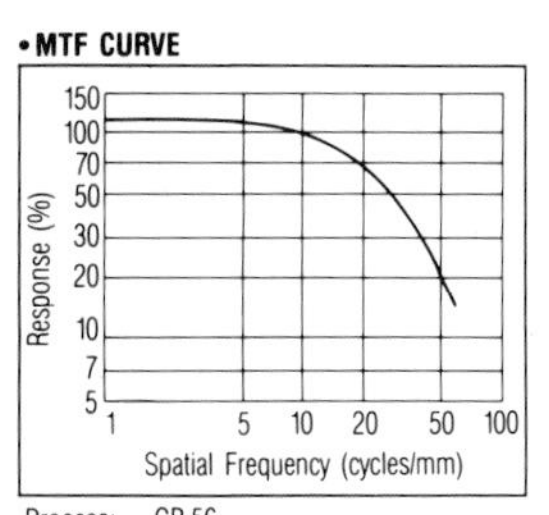

Process: CR-56
Illuminant: Daylight

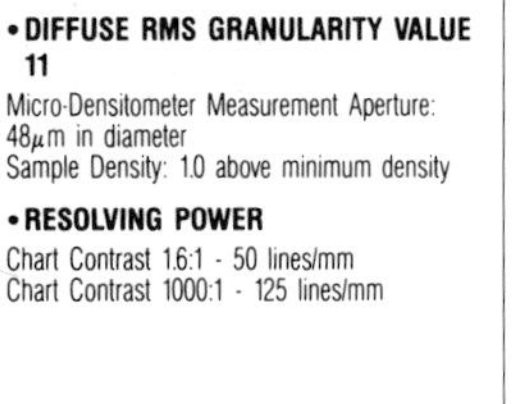

• DIFFUSE RMS GRANULARITY VALUE 11

Micro-Densitometer Measurement Aperture: 48μm in diameter
Sample Density: 1.0 above minimum density

• RESOLVING POWER

Chart Contrast 1.6:1 - 50 lines/mm
Chart Contrast 1000:1 - 125 lines/mm

FUJICHROME 400 Professional D [RHP] ISO 400/27°, Daylight Type

SPECTRAL SENSITIVITY CURVES

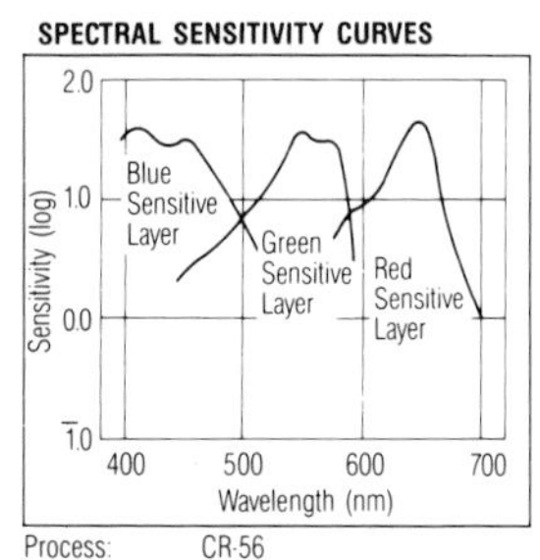

Process: CR-56
Densitometry: Status A
Density: 1.0 above minimum density
Sensitivity equals reciprocal of exposure (ergs/cm²) required to produce specified density

• SPECTRAL DYE DENSITY CURVES

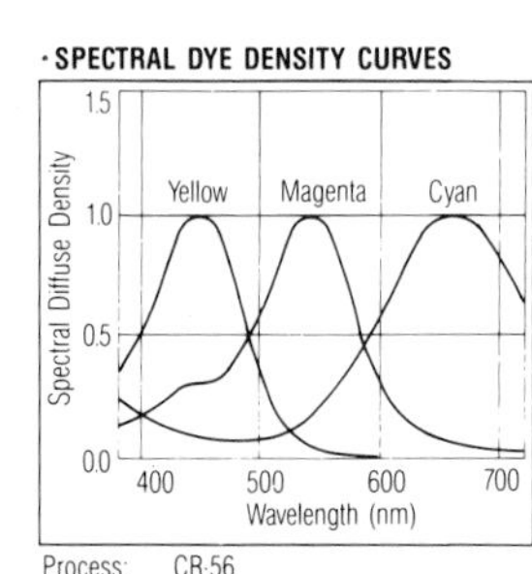

Process: CR-56
Exposure: Separated Light

• CHARACTERISTIC CURVES

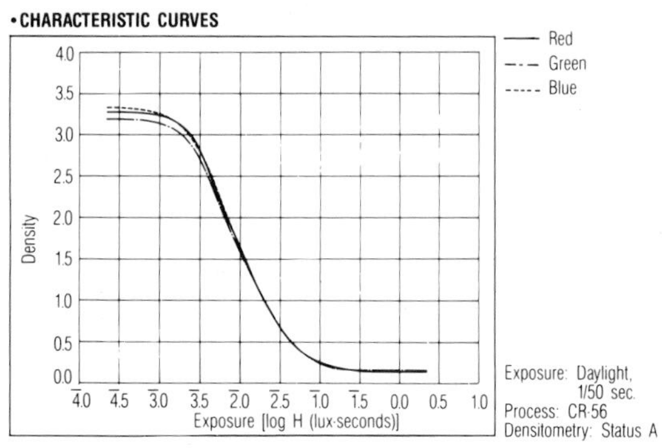

Exposure: Daylight, 1/50 sec.
Process: CR-56
Densitometry: Status A

• MTF CURVE

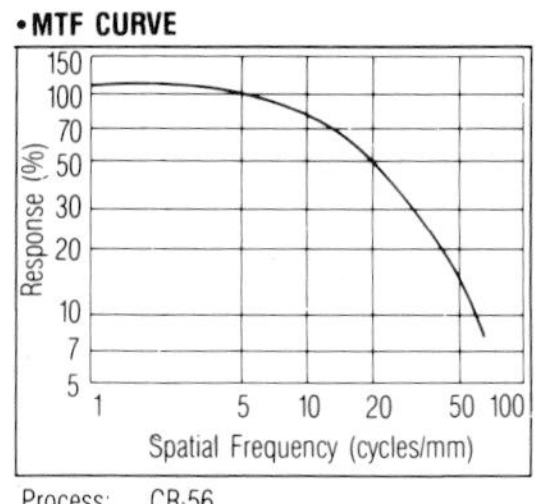

Process: CR-56
Illuminant: Daylight

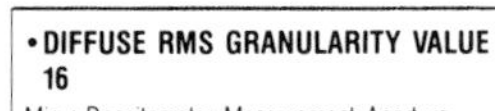

• DIFFUSE RMS GRANULARITY VALUE 16

Micro-Densitometer Measurement Aperture: 48μm in diameter
Sample Density: 1.0 above minimum density

• RESOLVING POWER

Chart Contrast 1.6:1 - 40 lines/mm
Chart Contrast 1000:1 - 125 lines/mm

Kodak Ektachrome 200 Professional Film

DIFFUSE RMS GRANULARITY VALUE: 13

(Read at a gross diffuse density of 1.0, using a 48-micrometre aperture, equivalent viewing magnification, 12X.)

KODAK EKTACHROME 200
Professional Film 5036 (Daylight)

CHARACTERISTIC CURVES

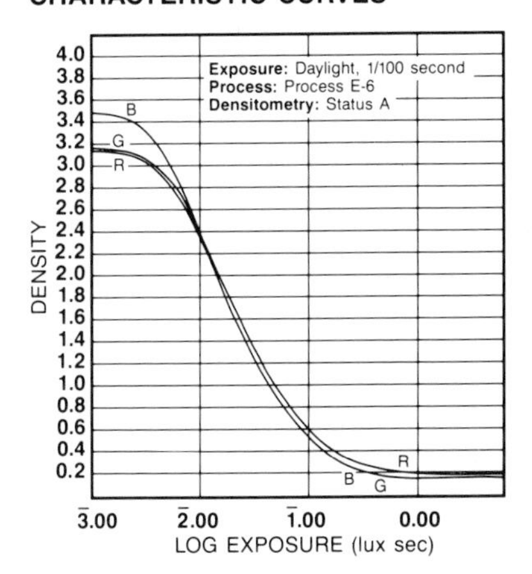

KODAK EKTACHROME 200
Professional Film 5036 (Daylight)

SPECTRAL SENSITIVITY CURVES

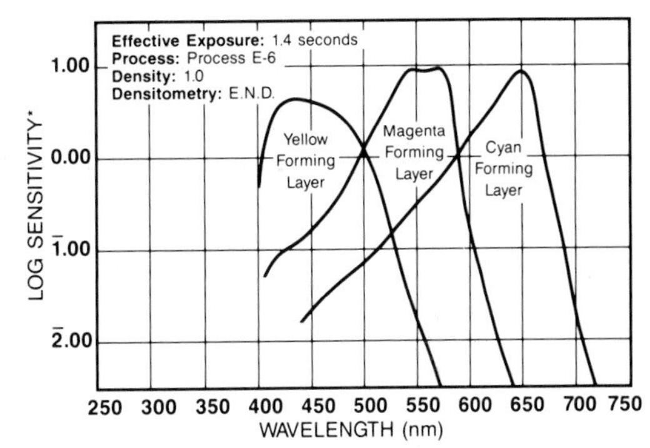

*Sensitivity = reciprocal of exposure (ergs/cm²) required to produce specified density

KODAK EKTACHROME 200
Professional Film 5036 (Daylight)

MODULATION TRANSFER CURVE

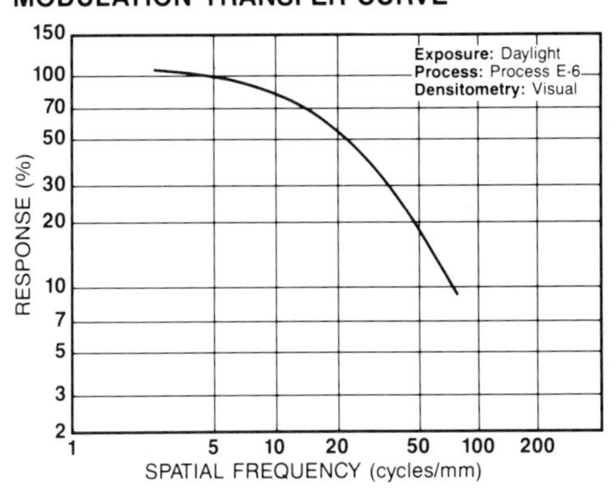

KODAK EKTACHROME 200
Professional Film 5036 (Daylight)

SPECTRAL DYE DENSITY CURVES

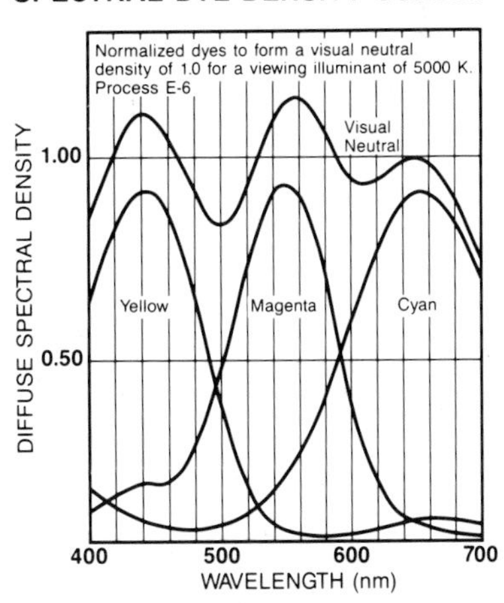

RESOLVING POWER VALUES:

Test-Object Contrast 1.6:1 (ISO RPL)	50 lines per mm
Test-Object Contrast 1000:1 (ISO RP)	100 lines per mm

Kodak Ektachrome 100 HC Film

IMAGE-STRUCTURE CHARACTERISTICS

Diffuse rms Granularity* 11

Resolving Power†	TOC 1.6:1	50 lines/mm
	TOC 1000:1	100 lines/mm

*Read at a gross diffuse visual density of 1.0, using a 48-micrometre aperture, 12X magnification.

†Determined according to a method similar to the one described in ISO 6328, *Photography—Photographic Materials—Determination of ISO Resolving Power.*

CHARACTERISTIC CURVES

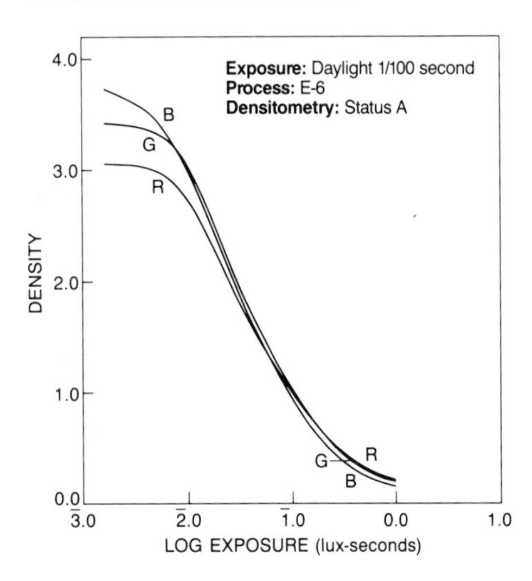

MODULATION-TRANSFER CURVE

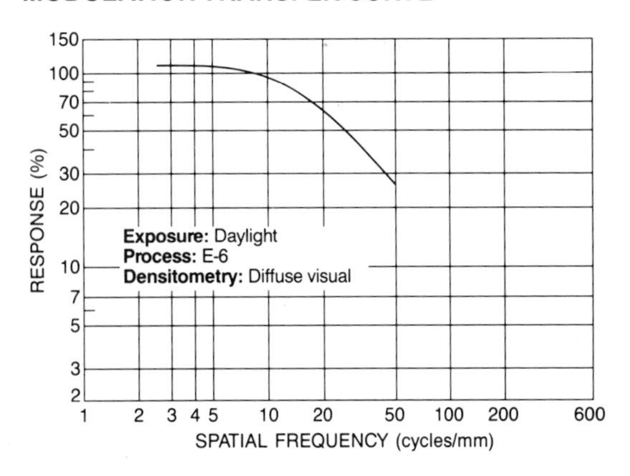

SPECTRAL-SENSITIVITY CURVES

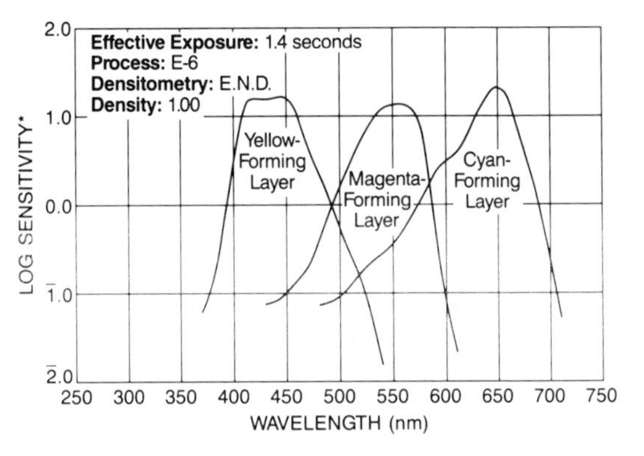

*Sensitivity = reciprocal of exposure (ergs/cm²) required to produce specified density

SPECTRAL-DYE-DENSITY CURVES

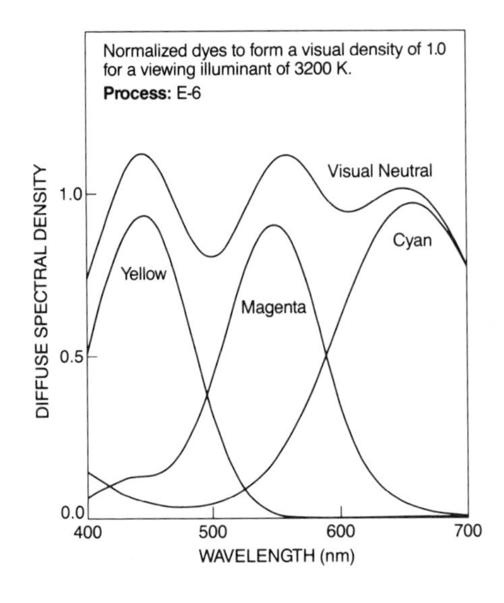

NOTICE: The sensitometric curves and data in this publication represent product tested under the conditions of exposure and processing specified. They are representative of production coatings, and therefore do not apply directly to a particular box or roll of photographic material. They do not represent standards or specifications that must be met by Eastman Kodak Company. The company reserves the right to change and improve product characteristics at any time.

Kodak T-Max 100 and 400 Professional Films

DARKROOM RECOMMENDATIONS

Do not use a safelight. Handle unprocessed film in total darkness. *Do not* develop these films by inspection.

Note: The afterglow (fluorescence) from some types of white light may fog these films. Make sure your darkroom is *completely* dark before you handle unprocessed film.

STORAGE AND HANDLING

Store unexposed film at 75 °F (24 °C) or lower in the original sealed package. For protection from heat in areas with temperatures consistently higher than 75 °F (24 °C), you can store the film in a refrigerator. If film has been refrigerated, allow the package to warm up to room temperature for 2 to 3 hours before opening it.

Load and unload roll-film cameras in subdued light, and rewind the film completely before unloading the camera. For best results, process the film promptly after exposure.

Total darkness is required when you remove the film from the magazine or load and unload film holders.

Store processed film in a cool, dry place.

EXPOSURE

The speed numbers for these films are expressed as Exposure Indexes (EI). Use these exposure indexes with meters or cameras marked for ISO/ASA or ISO °/DIN speeds in daylight or artificial light.

The developer you use to process these films affects the exposure index. Set your camera or meter (marked for ISO/ASA or ISO °/DIN speeds) at the speed for your developer given in the table.

Exposure Index (EI)		
KODAK Developer or Developer and Replenisher	**T-MAX 100 Professional Film**	**T-MAX 400 Professional Film**
T-MAX	**100/21°**	**400/27°**
T-MAX RS	**100/21°**	**400/27°**
D-76	**100/21°**	**400/27°**
D-76 (1:1)	**100/21°**	**400/27°**
HC-110 (Dil B)	100/21°	320/26°
MICRODOL-X	50/18°	200/24°
MICRODOL-X (1:3)	100/21°	320/26°
DURAFLO RT	80/20°	400/27°

Note: The developers and exposure indexes in **bold type** are the primary recommendations.

Under most conditions, you'll obtain highest quality with normal exposure at the rated exposure index and normal development. For high-contrast scenes, you'll obtain highest quality if you increase exposure by one or two stops and process the film normally.

If normal development produces negatives that are consistently too low in contrast, increase the development time slightly (10 to 15 percent). If negatives are too contrasty, decrease the development time slightly (10 to 15 percent). See "Adjusting Film Contrast" on page 12.

If your negatives are too thin, increase exposure by using a lower exposure index; if too dense, reduce exposure by using a higher exposure index.

Spectral-Sensitivity Curves

The blue sensitivity of KODAK T-MAX 100 and 400 Professional Films is slightly less than that of other Kodak black-and-white films. This enables the response of these films to be closer to the response of the human eye. Therefore, blues may be recorded as slightly darker tones with these films—a more natural rendition.

KODAK T-MAX 100 Professional Film / 5052

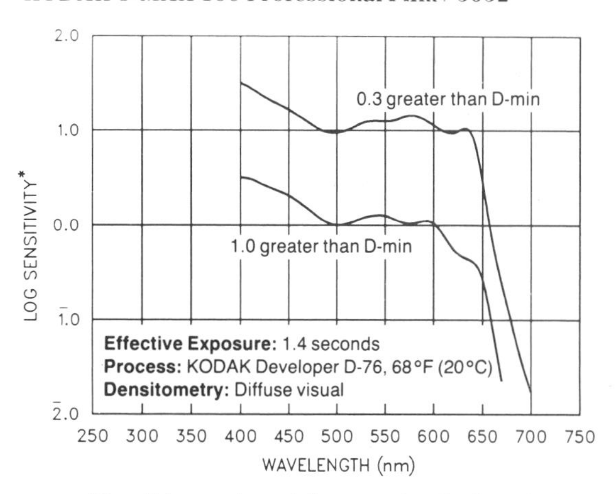

*Sensitivity = reciprocal of exposure (ergs/cm²) required to produce specified density

KODAK T-MAX 400 Professional Film / 5053

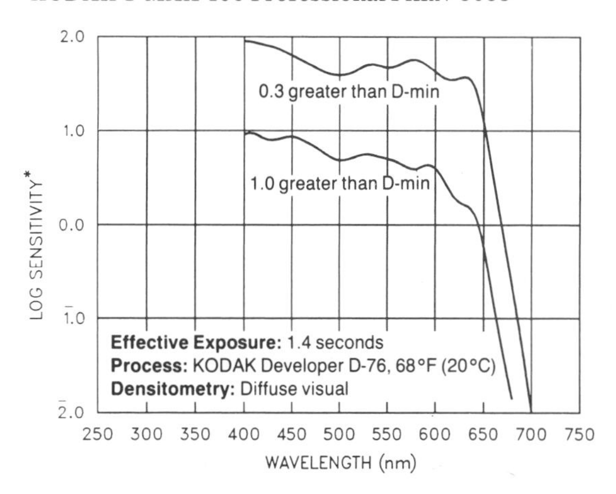

*Sensitivity = reciprocal of exposure (ergs/cm²) required to produce specified density

Kodak T-Max 100 and 400 Professional Films

MANUAL AND ROTARY-TUBE PROCESSING

Handle unprocessed film in total darkness.

These starting-point recommendations are intended to produce a contrast index of 0.56 with T-MAX 100 Professional Film and 0.60 with T-MAX 400 Professional Film. These development times will produce negatives with a contrast appropriate for printing with a diffusion enlarger. To adjust contrast for printing with a condenser enlarger, see "Adjusting Film Contrast" on page 12. Tank-development times shorter than 5 minutes may produce unsatisfactory uniformity.

Small-Tank Processing (8- or 16-ounce tank)—Rolls

Agitate once per 30 seconds.

With small single- or double-reel tanks, drop the loaded film reel into the developer and attach the top to the tank. Firmly tap the tank on the top of the work surface to dislodge any air bubbles. Provide initial agitation of 5 to 7 inversion cycles in 5 seconds, i.e., extend your arm and vigorously twist your wrist 180 degrees.

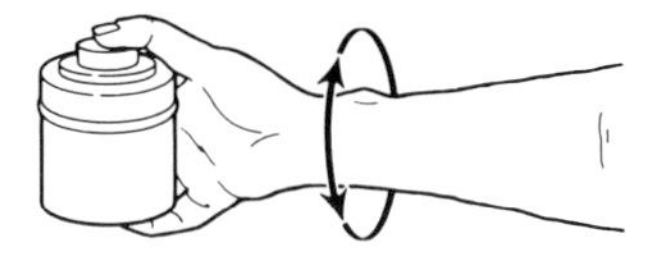

Then repeat this agitation procedure at 30-second intervals for the rest of the development time.

KODAK Developer or Developer and Replenisher	KODAK T-MAX 100 Professional Film					KODAK T-MAX 400 Professional Film				
	Development Time in Minutes					Development Time in Minutes				
	65°F (18°C)	68°F (20°C)	70°F (21°C)	72°F (22°C)	75°F (24°C)	65°F (18°C)	68°F (20°C)	70°F (21°C)	72°F (22°C)	75°F (24°C)
T-MAX	NR	8	7½	7	**6½**	NR	7	**6½**	6½	**6**
T-MAX RS	NR	8	7	7	**6**	NR	7	6	6	**5**
D-76	10½	**9**	8	7	6	9	**8**	7	6½	5½
D-76 (1:1)	14½	**12**	11	10	8½	14½	**12½**	11	10	9
HC-110 (Dil B)	8	**7**	6½	6	5	6½	**6**	5½	5	4½
MICRODOL-X	16	**13½**	12	10½	8½	12	**10½**	9	8½	7½
MICRODOL-X (1:3)	NR	NR	20	18½	**16**	NR	NR	20	18½	**16**
T-MAX (1:7)	—	—	—	—	**11½**	—	—	—	—	**10**
T-MAX (1:9)	—	—	—	—	**15**	—	—	—	—	**14**
T-MAX RS (1:7)	—	—	—	—	**8**	—	—	—	—	**7**
T-MAX RS (1:9)	—	—	—	—	**13½**	—	—	—	—	**12½**

NR = Not recommended

Note: The development times in **bold type** are the primary recommendations.

To prepare a modified dilution of T-MAX Developer (1:7 or 1:9), use the appropriate ratio of concentrate to water for the amount of solution you need.

To prepare a modified dilution of T-MAX RS Developer and Replenisher (1:7 or 1:9), first mix the normal working-strength solution from Parts A and B according to the instructions packaged with the chemicals. Then use the appropriate ratio of working-strength solution to water for the amount of solution you need. To prepare a 1:7 dilution, mix 5 parts of working-strength solution with 3 parts water. To prepare a 1:9 dilution, mix 1 part working-strength solution with 1 part water. We **do not** recommend using more dilute solutions of T-MAX RS Developer and Replenisher.

Rinse at 65 to 75°F (18 to 24°C) with agitation in KODAK Indicator Stop Bath or running water for 30 seconds.

Fix at 65 to 75°F (18 to 24°C) for 3 to 5 minutes with vigorous agitation in KODAK Rapid Fixer. Be sure to agitate the film frequently during fixing.

Note: To keep fixing times as short as possible, we strongly recommend using KODAK Rapid Fixer. If you use another fixer, such as KODAK Fixer or KODAFIX Solution, fix for 5 to 10 minutes or twice the time it takes for the film to clear. You can check the film for clearing after 3 minutes in KODAK Rapid Fixer or 5 minutes in KODAK Fixer or KODAFIX Solution.

Important: Your fixer will be exhausted more rapidly with these films than with other films. If your negatives show a magenta (pink) stain after fixing, your fixer may be near exhaustion, or you may not have used a long enough time. If the stain is slight, it will not affect negative contrast or printing times. If it is pronounced and irregular over the film surface, refix the film in fresh fixer.

Wash for 20 to 30 minutes in running water at 65 to 75°F (18 to 24°C) with a flow rate that provides at least one complete change of water in 5 minutes. You can wash long rolls on the processing reel. To save time and conserve water, use KODAK Hypo Clearing Agent.

Dry film in a dust-free place. To minimize drying marks, treat the film with KODAK PHOTO-FLO Solution after washing, or wipe the surfaces carefully with a KODAK Photo Chamois or soft viscose sponge.

Kodak T-Max 100 and 400 Professional Films

Characteristic Curves

KODAK T-MAX 100 Professional Film / 5052

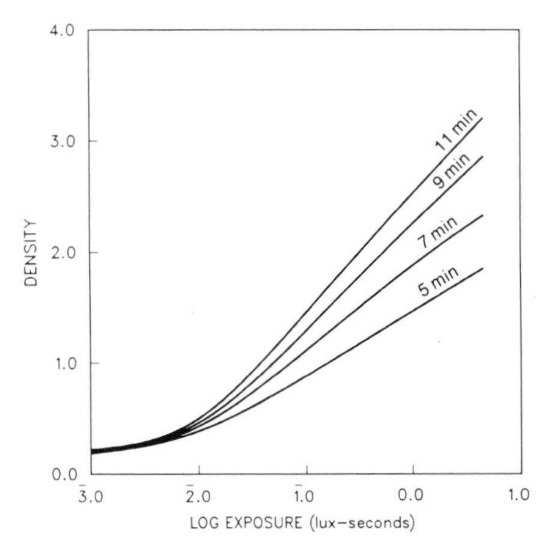

Exposure: Daylight
Process: Small tank, KODAK T-MAX Developer, 75 °F (24 °C)
Densitometry: Diffuse visual

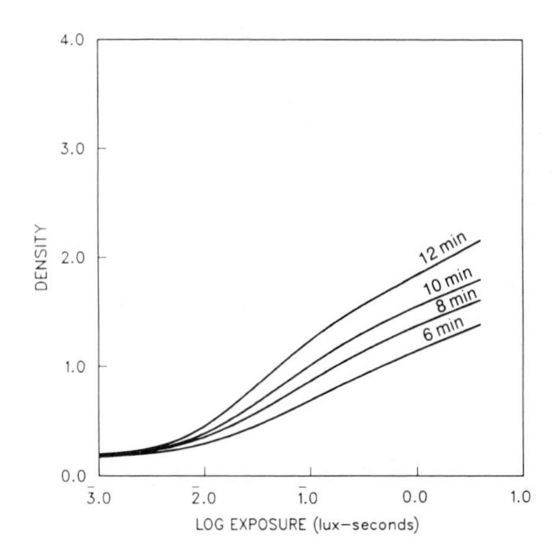

Exposure: Daylight
Process: Small tank, KODAK Developer D-76, 68 °F (20 °C)
Densitometry: Diffuse visual

KODAK T-MAX 400 Professional Film / 5053

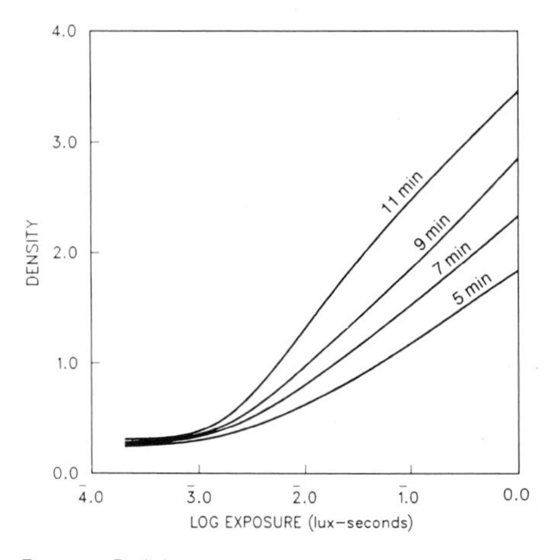

Exposure: Daylight
Process: Small tank, KODAK T-MAX Developer, 75 °F (24 °C)
Densitometry: Diffuse visual

Exposure: Daylight
Process: Small tank, KODAK Developer D-76, 68 °F (20 °C)
Densitometry: Diffuse visual

IMAGE-STRUCTURE CHARACTERISTICS

The data in this section are based on development at 68 °F (20 °C) in KODAK Developer D-76.

KODAK T-MAX 100 Professional Film

Diffuse rms Granularity* 8

Resolving Power†	TOC 1.6:1 TOC 1000:1	63 lines/mm 200 lines/mm

KODAK T-MAX 400 Professional Film

Diffuse rms Granularity* 10

Resolving Power†	TOC 1.6:1 TOC 1000:1	50 lines/mm 125 lines/mm

*Read at a net diffuse density of 1.00, using a 48-micrometre aperture, 12X magnification.

†Determined according to a method similar to the one described in ISO 6328, *Photography—Determination of ISO Resolving Power.*

Kodak T-Max P3200 Professional Film

DARKROOM RECOMMENDATIONS

Do not use a safelight. Handle unprocessed film in total darkness. **Do not** develop this film by inspection.

Note: Some darkroom timers will glow (fluoresce) for some time after you turn off the lights in a darkroom. To avoid fogging this film, turn the face of timers away from the area where you handle unprocessed film.

The afterglow (fluorescence) from some types of white light will also fog this film. Make sure your darkroom is *completely* dark before you handle unprocessed film.

STORAGE AND HANDLING

KODAK T-MAX P3200 Professional Film is very sensitive to environmental radiation; expose and process it promptly. Request *visual* inspection of this film at airport x-ray inspection stations.

Store unexposed film at 75 °F (24 °C) or lower in the original sealed package. For protection from heat in areas with temperatures consistently higher than 75 °F (24 °C), you can store the film in a refrigerator. If film has been refrigerated, allow the package to warm up to room temperature for 2 to 3 hours before opening it.

Load and unload your camera in subdued light, and rewind the film completely before unloading the camera.

Store processed film in a cool, dry place.

EXPOSURE

KODAK T-MAX P3200 Professional Film is specially designed to be used as *multi-speed* film. The speed you use depends on your application; make tests to determine the appropriate speed.

The nominal speed is EI 1000 when the film is processed in KODAK T-MAX Developer or KODAK T-MAX RS Developer and Replenisher, or EI 800 when it is processed in other Kodak black-and-white developers. It was determined in a manner published in ISO standards. For ease in calculating exposure and for consistency with the commonly used scale of film-speed numbers, the nominal speed has been rounded to EI 800.

Because of its great latitude, you can expose this film at EI 1600 and yield negatives of high quality. There will be no change in the grain of the final print, but there may be a slight loss of shadow detail. When you need higher speed, you can expose this film at EI 3200 or 6400. At these speeds, there will be a slight increase in contrast and granularity with additional loss of shadow detail. (See the processing tables for adjusted development times.)

IMAGE-STRUCTURE CHARACTERISTICS

The data in this section are based on development at 68 °F (20 °C) in KODAK Developer D-76.

Diffuse rms Granularity* 18

Resolving Power†	TOC 1.6:1 TOC 1000:1	40 lines/mm 125 lines/mm

*Read at a net diffuse density of 1.00, using a 48-micrometre aperture, 12X magnification.

†Determined according to a method similar to the one described in ISO 6328, *Photography—Determination of ISO Resolving Power.*

Spectral-Sensitivity Curves

KODAK T-MAX P3200 Professional Film / 5054

The blue sensitivity of KODAK T-MAX P3200 Professional Film is slightly less than that of other Kodak black-and-white films. This enables the response of this film to be closer to the response of the human eye. Therefore, blues may be recorded as slightly darker tones with this film—a more natural rendition.

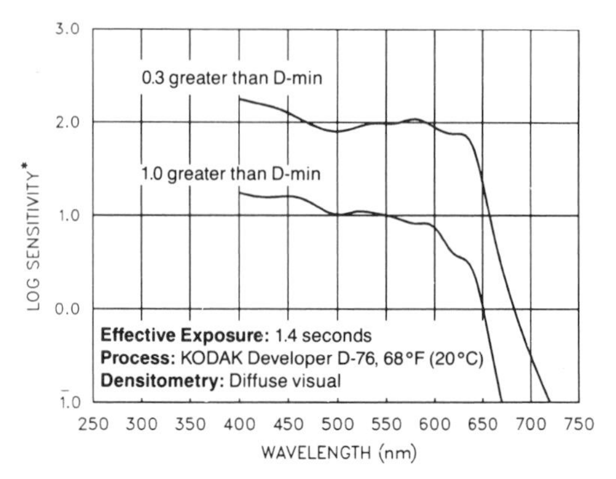

*Sensitivity = reciprocal of exposure (ergs/cm²) required to produce specified density

Characteristic Curves

KODAK T-MAX P3200 Professional Film / 5054

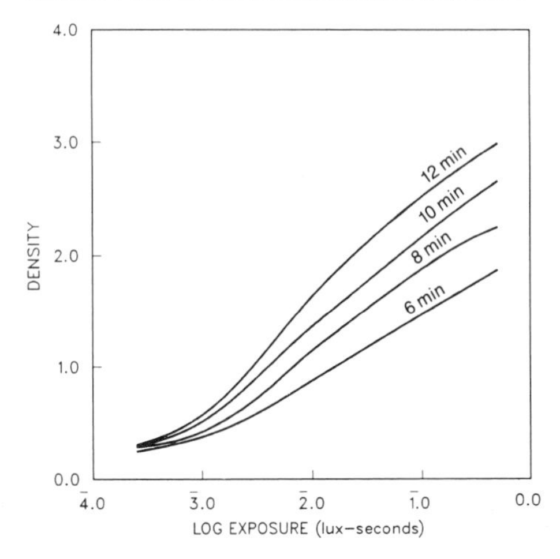

Exposure: Daylight
Process: Small tank, KODAK T-MAX Developer, 75 °F (24 °C)
Densitometry: Diffuse visual

Kodak T-Max P3200 Professional Film

Small-Tank Processing (8- or 16-ounce tank)

Agitate once per 30 seconds.

With small single- or double-reel tanks, drop the loaded film reel into the developer and attach the top to the tank. Firmly tap the tank on the top of the work surface to dislodge any air bubbles. Provide initial agitation of 5 to 7 inversion cycles in 5 seconds, i.e., extend your arm and vigorously twist your wrist 180 degrees as shown at the right.

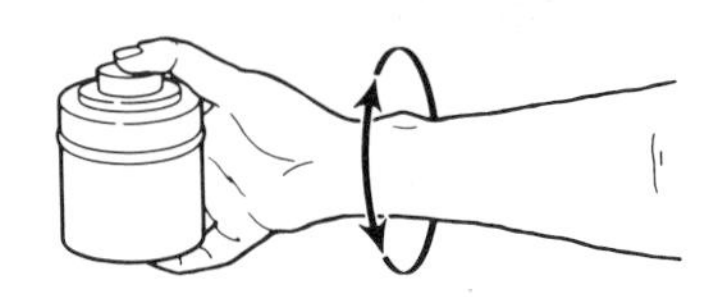

Then repeat this agitation procedure at 30-second intervals for the rest of the development time.

KODAK Developer or Developer and Replenisher	Exposed at EI	Development Time in Minutes					
		68°F (20°C)	70°F (21°C)	72°F (22°C)	75°F (24°C)	80°F (27°C)	85°F (29°C)
T-MAX	400/27°	7½	7	6½	6	5	4
	800/30°	8	7½	7	6½	5½	4½
	1600/33°	8½	8	7½	7	6	5
	3200/36°	11½	11	10½	9½	8	6½
	6400/39°	14	13	12	11	9½	8
	12,500/42°*	16	15½	14½	12½	10½	9
	25,000/45°*	NR	17½	16	14	12	10
T-MAX RS	400/27°	8	7	6½	6	5½	5
	800/30°	9	8½	7½	6½	6	5½
	1600/33°	10½	9½	8½	7½	7	6
	3200/36°	13	12	11	10	9	8
	6400/39°	15	14	13	11	10	9
	12,500/42°*	18	16	14	12	11	10
	25,000/45°*	NR	NR	16	14	13	11
D-76	400/27°	10½	9½	8½	7½	6	4½
	800/30°	11	10	9	8	6½	5
	1600/33°	11½	10½	9½	8½	7	5½
	3200/36°	15	13½	12½	11	8½	7½
	6400/39°	17½	16	14½	12½	10½	9
HC-110 (Dil B)	400/27°	7½	6½	5½	5	4½	3½
	800/30°	8	7	6	5½	4¾	4
	1600/33°	9	7½	6½	6	5	4½
	3200/36°	11½	10	8½	7½	6½	5¾
	6400/39°	14	12	10½	9½	8	6¾
T-MAX (1:7)	800/30°	—	—	—	12½	—	—
T-MAX (1:9)	800/30°	—	—	—	17	—	—
T-MAX RS (1:7)	800/30°	—	—	—	10	—	—
T-MAX RS (1:9)	800/30°	—	—	—	15	—	—

*Make tests to determine if results at these speeds are acceptable for your needs.

NR = Not recommended

Note: These development times are *starting-point* recommendations. Make tests to determine the best development time for your application.

To prepare a modified dilution of T-MAX Developer (1:7 or 1:9), use the appropriate ratio of concentrate to water for the amount of solution you need.

To prepare a modified dilution of T-MAX RS Developer and Replenisher (1:7 or 1:9), first mix the normal working-strength solution from Parts A and B according to the instructions packaged with the chemicals. Then use the appropriate ratio of working-strength solution to water for the amount of solution you need. To prepare a 1:7 dilution, mix 5 parts of working-strength solution with 3 parts water. To prepare a 1:9 dilution, mix 1 part working-strength solution with 1 part water. We **do not** recommend using more dilute solutions of T-MAX RS Developer and Replenisher.

Rinse, fix, and wash as for Kodak T-Max 100 and 400 Professional Films.

Kodak Technical Pan Films

DESCRIPTION

KODAK Technical Pan Film is a black-and-white panchromatic negative film with extended red sensitivity. The 2415 Film is available in 135 size and 35 mm long rolls; it has a dimensionally stable 4-mil ESTAR-AH Base. The 4415 Film is available in 4 x 5-inch and 8 x 10-inch sheets with a dimensionally stable 7-mil ESTAR Thick Base. The 6415 Film is available in 120 size with a 3.6-mil acetate base. All three films have good latent-image keeping, a dyed-gel backing to suppress halation and curl, and a built-in 0.1-density dye that suppresses light piping.

EXPOSURE

Since you can use Technical Pan Film under a wide range of exposure conditions, we give recommendations only for the common applications. Note in the table that the exposure index is a function of the processing conditions and the contrast produced. Compare these values with the characteristic curves on pages 9, 10, and 11 to choose the appropriate contrast and exposure index. Exposure-index values given here are for use with meters marked for ISO (ASA/DIN) speeds or exposure indexes and are starting points for trial exposures.

You should bracket exposures by whole-stop increments for initial tests; use half-stop incrememts for critical applications.

Exposure- and Contrast-Index Values for Various Development Conditions

Contrast Index	KODAK Developer	Development Time (minutes) at 68°F (20°C)	Exposure Index
High 2.50	DEKTOL	3	200
2.40–2.70	D-19 (1:2)	4–7	125–160
2.25-2.50	D-19	2-8	100-200
1.20–2.10	HC-110 (Dil B)	4–12	100–250
1.25–1.75	HC-110 (Dil D)	4–8	80–125
1.00–2.10	D-76	6–12	50–125
0.80–0.95	HC-110 (Dil F)	6–12	32–64
Low 0.50–0.70	TECHNIDOL Liquid	5–11	16–25

Tests show that Technical Pan Film is approximately 10 percent more sensitive to tungsten light than to daylight. You may also find an increase in contrast of approximately 5 percent with tungsten light.

Filter Factors

Multiply your exposure by the following filter factors when you use filters. If you use a through-the-lens meter, take the meter reading without the filter over the lens, and then calculate your exposure by using the filter factor. Where no filter factor is listed in the table, no test was made with that filter.

KODAK WRATTEN Gelatin Filter	Tungsten Filter Factor*	Daylight Filter Factor†
No. 8	1.2	1.5
No. 11	5	—
No. 12	1.2	—
No. 15	1.2	2.0
No. 25	2	3.0
No. 47	25	—
No. 58	12	—

*Based on a 1-second tungsten exposure with development for 8 minutes at 68°F (20°C) in KODAK HC-110 Developer (Dilution D).

†Based on a 1/25-second daylight exposure with development for 9 minutes at 68°F (20°C) in KODAK TECHNIDOL Liquid Developer.

Adjustments for Long and Short Exposures

No exposure adjustments are required for exposures between 1/10,000 and 1 second. However, for exposures shorter than 1/500 second, you may want to increase development to compensate for the lower contrast that very short exposures can produce.

If indicated Exposure Time Is (seconds)	Use This Lens-Aperture Adjustment	OR	This Exposure-Time Adjustment (seconds)	AND Use This Development Adjustment
1/10,000	None		None	+30%
1/1000	None		None	+20%
1/100	None		None	None
1/10	None		None	None
1	None		None	−10%
10	+½ stop		15	−10%
100	+1½ stop		NR	None

NR = Not recommended.

Changes in Speed and Contrast Due to Long- and Short-Exposure Adjustments

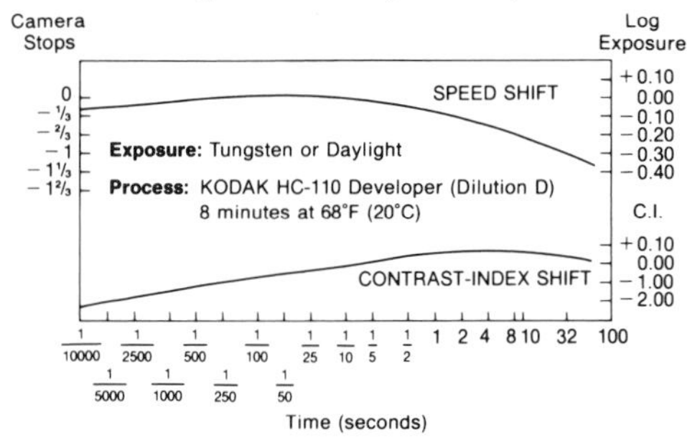

Kodak Technical Pan Films

PRINTING

The 0.1 neutral density built into the ESTAR-AH Base is one-half to one-third that found in conventional 35 mm picture-taking films. Correctly exposed and processed pictorial negatives may appear to be "thinner" than normal. It is important to take this into account when you judge the printability of negatives.

The micro-fine grain of Technical Pan Film makes possible printing at higher magnifications than are usually acceptable with conventional picture-taking films. Enlargements made at magnifications greater than 25X with highly specular (point-source) enlargers may show a random distribution of poorly defined white specks in otherwise dense areas. The specks are caused by tiny matte particles coated on the back surface of the film. You can mask the specks, with little loss in the overall sharpness of the image, by using an enlarger with a diffuse or semi-diffuse light source.

IMAGE-STRUCTURE CHARACTERISTICS

The data in this section are based on development at 68°F (20°C) in KODAK HC-110 Developer (Dilution D) for 8 minutes, KODAK TECHNIDOL Liquid Developer for 9 minutes.

	KODAK Developer	
	HC-110 (Dilution D)	TECHNIDOL Liquid
Diffuse RMS Granularity*	**8 (extremely fine)**	**5 (micro-fine)**
Resolving Power+ (lines per mm)		
TOC 1.6:1	125	100
TOC 1000:1	320	320

*Read at a net diffuse density of 1.0 using a 48-micrometre aperture and 12X magnification.

+Determined according to a method similar to the one described in ISO 6328-1982, *Method for Determining the Resolving Power of Photographic Materials*. The values given for this film represent extremely high resolving power.

SPECTRAL-SENSITIVITY CURVE

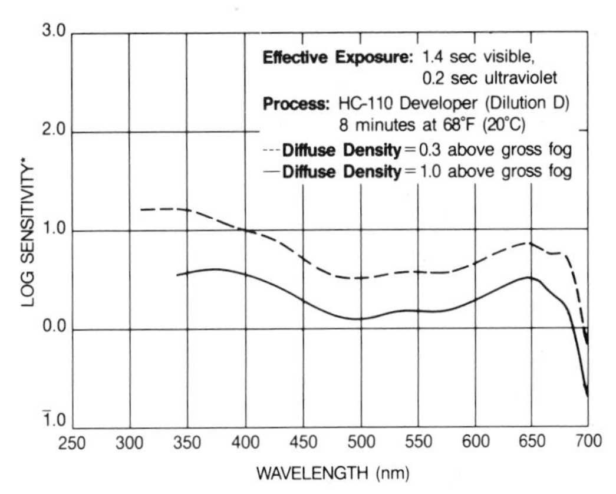

*Sensitivity = reciprocal of exposure (ergs/cm²) required to produce specified density

MODULATION-TRANSFER FUNCTION

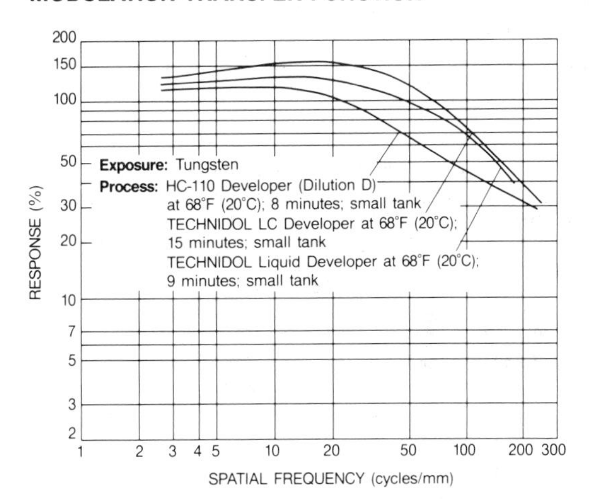

These photographic modulation-transfer values were determined using a method similar to that described in ANSI Standard PH-2.39-1977(R1984). The film was exposed with the specified illuminant to spatially varying sinusoidal test patterns with an aerial image modulation of a nominal 35 percent at the image plane, with processing as indicated. In most cases, these photographic modulation-transfer values are influenced by development adjacency effects, and are not equivalent to the true optical modulation-transfer function of the emulsion layer in the particular photographic product.

Note: KODAK TECHNIDOL LC Developer is no longer available. KODAK TECHNIDOL Liquid Developer is recommended as a substitute.

Kodak Technical Pan Films

CHARACTERISTIC CURVES*

*Densitometry: American Standard diffuse visual density

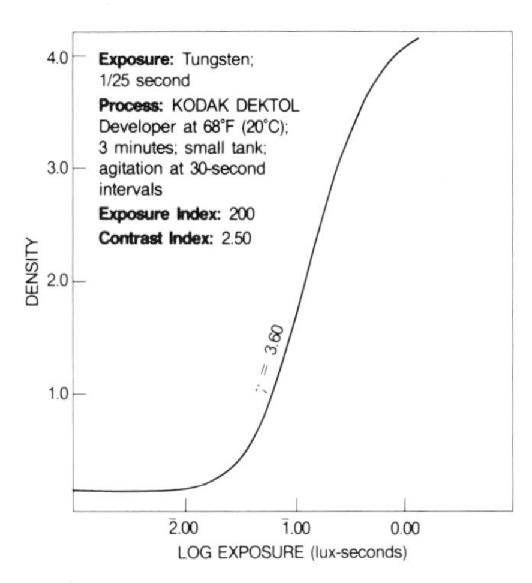

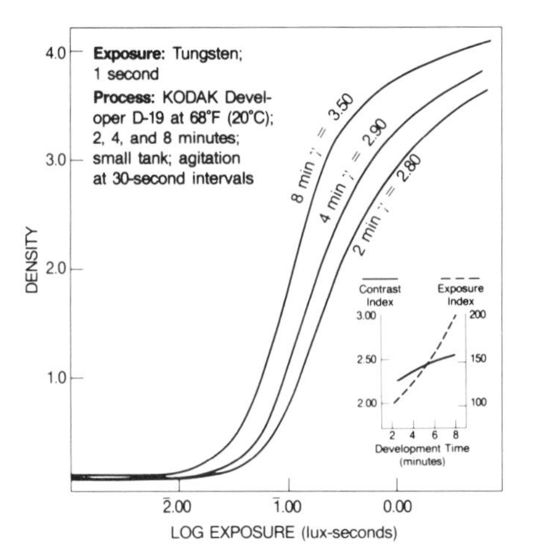

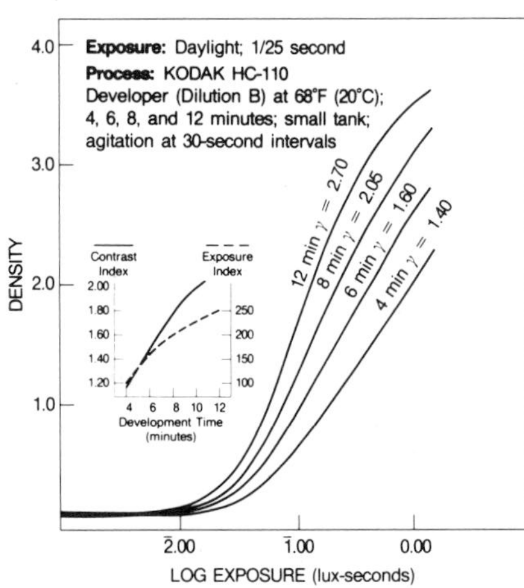

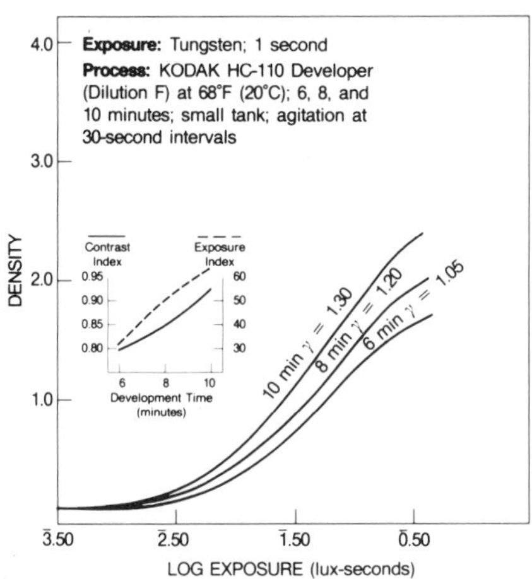

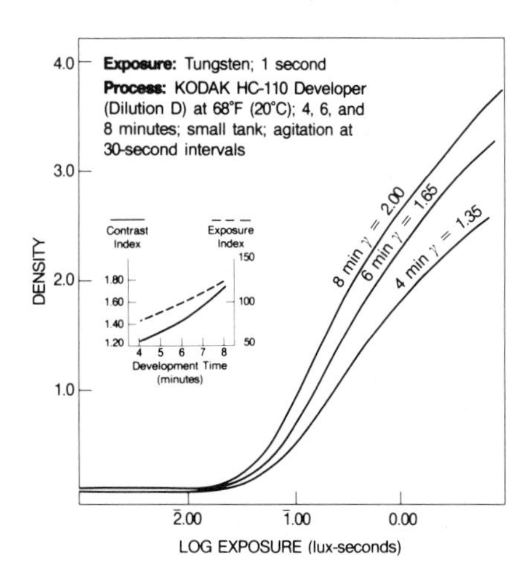

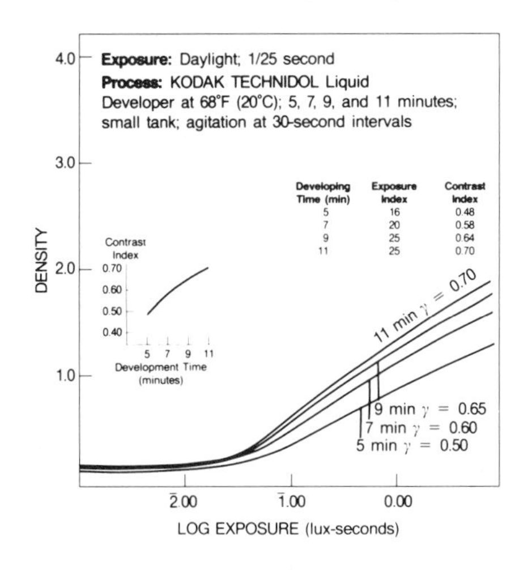

Index